Power Systems

For further volumes:
http://www.springer.com/series/4622

Jawad Faiz · Behzad Siahkolah

Electronic Tap-changer for Distribution Transformers

Springer

Professor Jawad Faiz
School of Electrical and Computer Engineering
University College of Engineering,
University of Tehran
Tehran
Iran
e-mail: jfaiz@ut.ac.ir

Dr. Behzad Siahkolah
Kavandish Company
Tehran
Iran
e-mail: behzad@kavandish.com

ISSN 1612-1287 e-ISSN 1860-4676

ISBN 978-3-642-19910-3 e-ISBN 978-3-642-19911-0

DOI 10.1007/978-3-642-19911-0

Springer Heidelberg Dordrecht London New York

Cover design: eStudio Calamar, Berlin/Figueres

Printed on acid-free paper

Springer is part of Springer Science+Business Media (www.springer.com)

Preface

Although many papers dealing with electronic tap-changer has been published during the last decades, there is not any comprehensive reference book, dealing with electronic tap-changer in its all aspects in bookshelves. This book is an endeavor to collect academic writings in this field and to present them in a comprehensive and orderly manner. It is useful to electrical engineers in industries and postgraduate and graduate students of electrical engineers.

Using a logical and systematical manner, designing and optimizing the full-electronic tap-changer for distribution transformers are carried. The design process is divided into two parts:

1. Designing tapped windings structure and switches configurations (design of power section)
2. Designing closed-loop control system

In the power section design of the full-electronic tap-changers a fully new perspective of all possible structures are introduced and categorized comprehensively. Since one of the major barriers in the expanding the application of the full-electronic tap-changer is its high cost, the major factors influencing the cost—number and ratings of the semi-conductor switches and the number of transformer taps—the cost function as an objective function is proposed and the optimal structure is then determined based on this objective function.

In the control section design, the mechanical tap-changer control system is reviewed; the full electronic tap-changer is modeled and closed-loop control of the full-electronic tap-changer is designed using a new method. In this method, first problem is made less complex by imposing five disciplined assumptions on the design process and design is carried out. These assumptions are then concealed one by one, the necessary modifications are applied in the control system, and the final design is followed.

Simulink software is used to simulate the power section of the full-electronic tap-changer with its control section. Simulation results show the correctness of the design procedure. Following the design of the full-electronic tap-changer, performance and capabilities of this system is investigated for power quality enhancement. It is compared with other custom power tools from economical outlook. Finally, a low-power 5 kW full-electronic tap-changer prototype is built and tested to show the whole design and simulation stages. Experimental results demonstrate the theoretical predictions.

Tehran, January 2011

Jawad Faiz
Behzad Siahkolah

Acknowledgments

This book would never have been possible without the help of a number of peoples and organizations over the last 10 years. We would especially like to thank the University of Tehran for financial support of this work. We would also like to thank Dr. Keyvanpanah, the Assistant Professor of the Department of English Language, Faculty of Foreign Languages of the University of Tehran for editing the text, M.R. Okhovvat for developing the required tap-changer control softwares, Kavandish Company in Tehran, Iran, for helping to build the prototype tap-changer.

Acknowledgments

Contents

Chapter 1
Past and Present

1.1 Introduction

Distribution transformer tap is a connection which is taken out from a node located between two ends of a winding [1]. This permits changes in voltage, current or turns ratio of the transformer after it has left the factory. The reasons to have a series of taps in the transformer are as follows [2]:

a. To fix the secondary voltage against the primary voltage changes;
b. To change the secondary voltage;
c. To provide an auxiliary secondary voltage for a specific application such as lighting;
d. To reduce voltage for starting rotating motors;
e. To provide a natural point for earthing or conducting unbalanced current in three-wire single-phase circuits or four-wire three-phase circuits.

In the transformers used in power systems the main reason for taps application is adjusting and controlling the voltage [3]. The load fluctuations change the voltage of the power system. It is noted that sometimes taps in power transformers are used to shift the phase angle [4, 5].

Tap-changers are categorized into two main groups [6]:

a. Off-circuit or no-load tap-changers;
b. On-load or under-load tap-changers.

Tap-changer with the ability to change taps while power is on called under-load tap-changer. If a tap-changer is built as such that its fixing requires its being disconnected from the power line, the tap-changer is called no-load tap-changer.

If there is a need to change the turns ratio over a long interval (for instance seasonal), the no-load tap-changer is used. Normally in the no-load tap-changers taps are changed manually by means of a selector outside the transformer tank. This selector may move linearly or circularly [2]. Figure 1.1 presents the structure

J. Faiz and B. Siahkolah, *Electronic Tap-changer for Distribution Transformers*, Power Systems, DOI: 10.1007/978-3-642-19911-0_1,

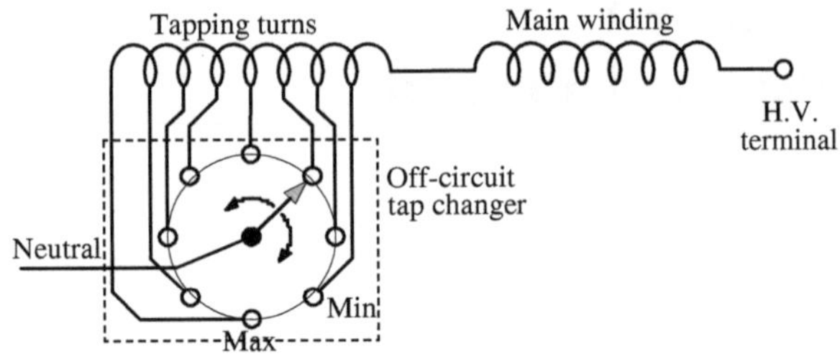

Fig. 1.1 Structure of an no-load tap-changer [7]

of a no-load tap-changer with a circular selector. For short intervals (different hours of day), it is not possible to disconnect the transformer from the power system and therefore under-load tap-changers must be used to change the taps. This type of tap-changer is widely used in power systems and their details are presented in the following parts.

1.2 Under-Load Tap-Changer

Before establishing large and wide transmission power networks, the common technique for voltage adjustment in small and isolated networks was the use of a generator. This voltage adjustment was done through by generator excitation system [8]. However, the voltage of different parts of a large network cannot be adjusted using such a technique. In larger networks an on-load tap-changer must be used in order to adjust the under-load voltage continuously. These tap-changers are fixed on the transmission transformers and also the main load supplying transformers in order to compensate voltage changes which are caused by load fluctuations [4].

The structure of under-load tap-changers is more complex than that of the no-load version. There are two basic features shared by all under-load tap-changer [2, 9, 10].

a. During tap changing stages, taps are never short-circuited; thus an impedance is used to prevent the short-circuit.
b. Load is not disconnected during tap-changing process. To this ends, there is typically a two-element circuit in the under-load tap-changer, when switching and tap changing are carried out in one element of the circuit, the load current continues to pass in the other element.

The impedance used in under-load tap-changers is a tapped resistor or inductor and so it is called resistor or inductor tap-changer. Although the advantage of resistors because of their short time duration and consequently longer life of contacts was already known, the earlier designs of tap-changers followed the tapped inductor version. The reason for the wide spread use of tapped-inductor under-load tap-changers was its ability to continuously pass the load current; while resistors (based on their rating) can pass the load current over a very short period of time. The absence this feature is the possible damage of the moving mechanical system of the tap-changers which stops the tap-changers operation, while the load current passes the tap-changer impedance. As mechanical systems used in the under-load tap-changers became advanced inductor tap-changers have gradually been replaced by resistor tap-changers [2].

1.2.1 Under-Load Reactor Tap-Changers

The reactor under-load tap-changer was invented in 1926 [5]. There are different circuits as used in this tap-changer. Figure 1.2 shows the simplest form of an under-load reactor tap-changer in which an under-load switch is connected to each tap. The under-load switch is a switch which can be turned on and off under load.

According to Fig. 1.2, the switches are connected to each other alternatively and form two distinctive groups. Each group is connected to a terminal of the reactor. Table 1.1 summarizes various positions of the tap-changer based on the different conditions of the switches.

According to Table 1.1, only switch no. 1 at position 1 is on and the current passes only an half of the reactor. In order to transfer the tap to position no. 2, in addition to switch no. 1, switch no. 2 will be on. In this case the reactor is placed between tap no. 1 and tap no. 2 and prevents the short circuit of the taps. Meanwhile, the load current continues through the middle terminal of the reactor, and the load voltage is the mean voltage of tap no. 1 and tap no. 2. Therefore, in this configuration we can conveniently get the taps mid-voltage and this is one of the advantages of this structure. To change the position to position no. 3, switch no. 1 is off and only switch no. 2 is on; in this case the second half of the reactor will be the load current path.

Similarly, tap changing to further positions are possible based on Table 1.1.

There are $(2n - 1)$ positions for the load voltage in the tap-changer of Fig. 1.2 (n is the number of taps) and this is an advantage. However, the number of under-load switches is equal to the number of taps and this increases the volume, cost and complexity of the system. Figure 1.3 presents an alternative under-load tap-changer in which there are two off-load selector switches to select the tap and only two under-load switches (diverter switch) exist for diverting the load current during tap-changing stage [2].

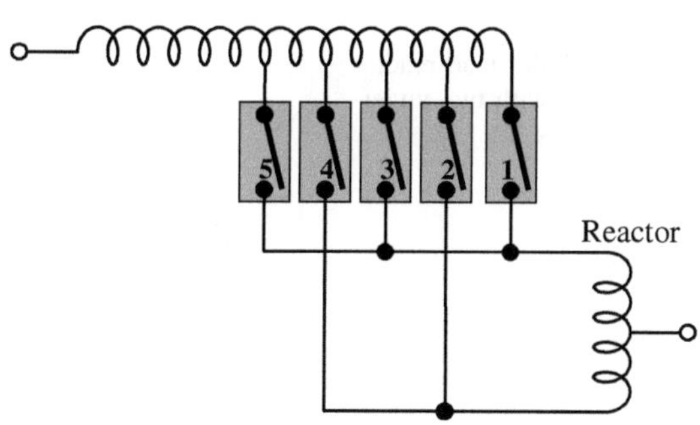

Fig. 1.2 The simplest form of an under-load reactor tap-changer [2]

Table 1.1 Stages of tap changing in tap-changer (Fig. 1.2)

Switch no./position no.	1	2	3	4	5	6	7	8	9
1	ON	ON							
2		ON	ON	ON					
3				ON	ON	ON			
4						ON	ON	ON	
5								ON	ON

Selector switches and diverter switches are coupled with the gears of tap-changing mechanical system. In this tap-changer, taps are changed in the following ways:

a. Diverter switch no. 1 is opened.
b. Selector switch no. 1 moves.
c. Diverter switch no. 1 is reclosed.
d. Diverter switch no. 2 is opened.
e. Selector switch no. 2 moves.
f. Diverter switch no. 2 is reclosed.

One other advantages of the tap-changer in Fig. 1.3 is that the voltage drop on the reactor is zero, except at the tap-changing time. The inductance must not be very large in order to decrease the voltage drop, but must be very small in order to increase the rotating current between the adjacent tapes during tap-changing stages.

As already mentioned, the main advantage of under-load reactor tap-changers is that the reactor stands the continuous load current, so if the moving mechanical system for the tap-changer is damaged and stopped during the tap-changing process, the tap-changer will not burn and transformer will not fail. On the other hand, under-load reactor tap-changers have also some drawbacks, such as low changing speed, long arcs during tap-changing, short life of contacts and large volume of reactors. This is the reason that the use of these tap-changers has become obsolete. However, some improvements have been made in their structure and they are used in North America.

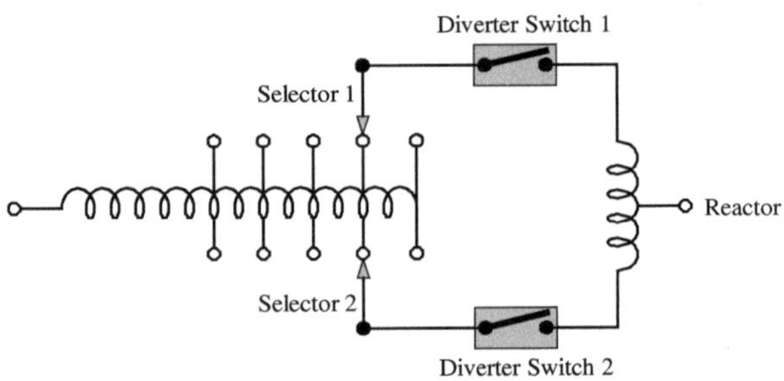

Fig. 1.3 Under-load reactor tap-changer with two under-load switches [2]

1.2.2 Under-Load Resistor Tap-Changers

Under-load reactor tap-changers have been almost totally replaced by fast under-load resistor tap-changers. Generally two basic circuit arrangements have been used in resistor tap-changers [2]:

a. Those which do tap selection and load current deviation in one single contact (type I).
b. Those which have distinctive diverter switches and tap selection switches (type II).

Figure 1.4 shows the different circuit arrangements of switches in the resistor tap-changers and Fig. 1.4a the type I arrangement. Such an arrangement is normally done by rotating switches and is called Pennant cycle. Figure 1.5 shows the stages of tap changing in this type of tap-changer. Figure 1.4b–d present the type II arrangements. In these figures the arrangement of the selector switches, similar to the diverter switches, differ from the type I arrangement. These arrangements are called the Flag cycle.

In large transformers the arrangement type II are normally used. Figure 1.6 presents the stages of tap-changing in Fig. 1.4d structure.

There are different structures for selector switches and transformer taps. Figure 1.7 shows three types of these structures [11].

Figure 1.7a shows a structure in which the taps are placed in linear steps. When the range of taps is wide, the structures of Fig. 1.7b and c can be use, in order to reduce the number of taps and their voltage between any one tap and the next. Structure of Fig. 1.7b shows a condition in which the taps winding can be inversed and, therefore, it will be possible to add to or subtract from the taps winding voltage from the main winding.

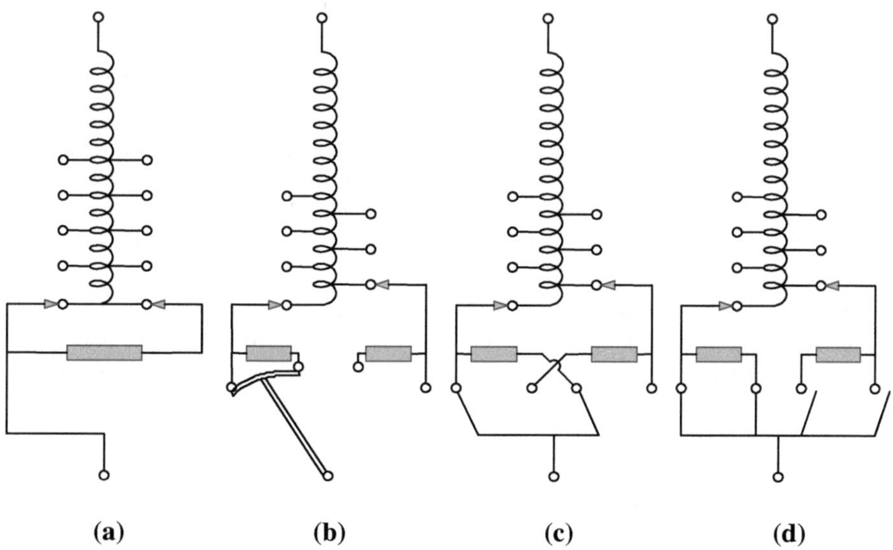

Fig. 1.4 Different arrangements of switches in resistor tap-changers

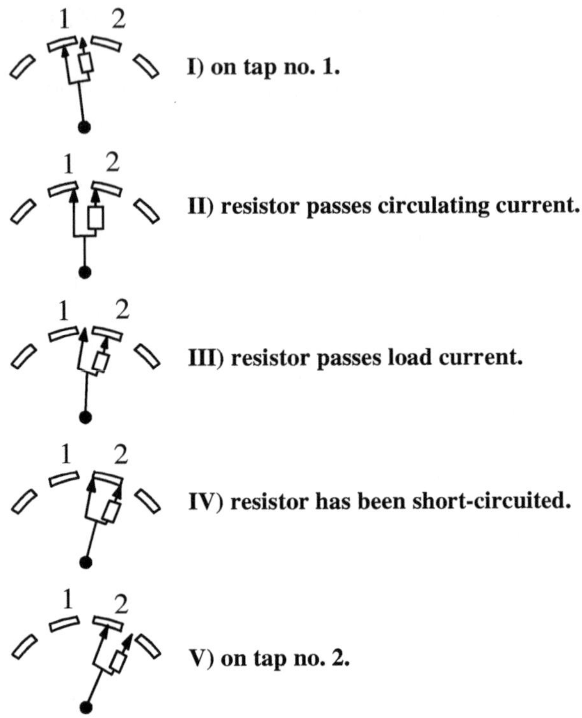

Fig. 1.5 Stages of tap-changing from tap number 1 to tap number 2 in the arrangement (Fig. 1.4a)

For precise adjustment, there are two types of taps: the taps with large steps and taps with small steps.

1.3 Limitations: Mechanical Under-Load Tap-Changers

In spite of advancement in the structure of mechanical under-load tap-changers, these tap-changers have some drawbacks; for example, the major factor which causes damage to power transformers is their tap-changer failure [12]. Some drawbacks of mechanical under-load tap-changers are as follows [13].

a. Contact Arc in Diverter Switches During Tap-Changing Process
An arc appears in the contacts of diverter switches at the time of make and breaks the load current. This arc causes impurity of the oil surrounding the diverter switches and wearing out of the contacts of the switches.

b. High Maintenance and Service Cost
Conditions of oil, contacts and mechanical parts of the mechanical under-load tap-changers must be inspected regularly. This is required due to arc and wearing-out of the moving mechanical parts of tap-changer.

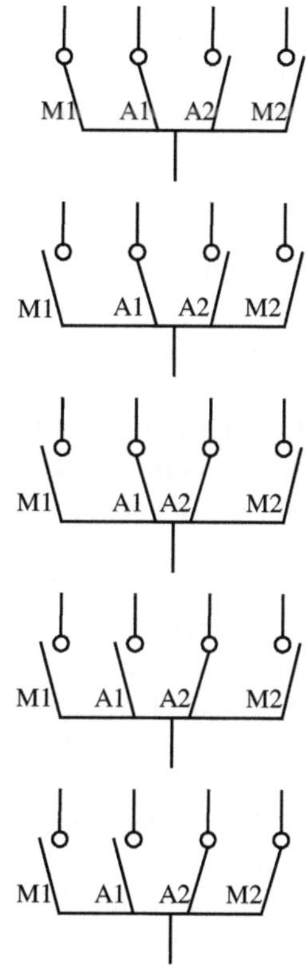

Fig. 1.6 Stages of tap-changing in structure (Fig. 1.4d) [2]

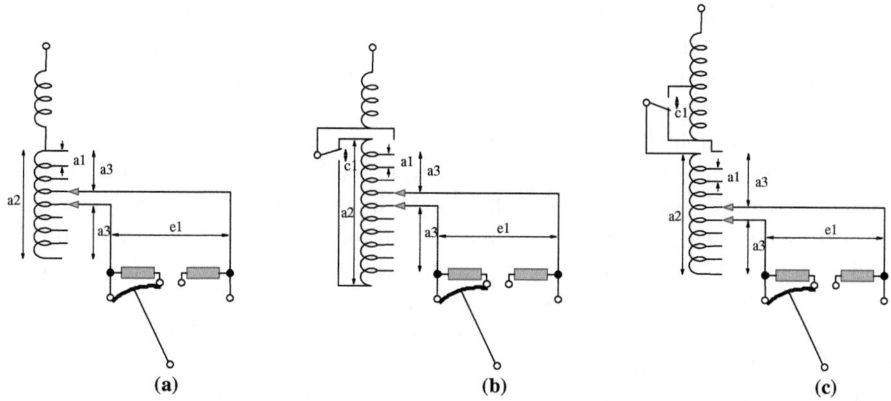

Fig. 1.7 Different structures for taps of selector switches in under-load resistor tap-changer [11]

c. Low Speed of Tap-Changing
The low speed of tap-changing originates from mechanical nature of tap-changing process and the required time for storing the desirable energy for tap-changing process. In an under-load resistor tap-changer, it takes 100 s to change tap number 1 to tap number 19 [14].

d. High Losses of Tap-Changer During Tap-Changing
This happens for under-load resistor tap-changer and reason is the use of passing resistors in this type of tap-changer.

In order to remove the above-mentioned limitations and drawbacks, the following new circuits and structures have been suggested for under-load tap-changers [15]. These are categorized into two major groups [15, 16].

a. Electronically Assisted Under-Load Tap-Changers (or Hybrid On-Load Tap-Changer)
In these tap-changers solid-state power switches have been used beside the mechanical switches in order to reduce the arc caused by tap-changing. Mechanical parts of the conventional under-load tap-changer systems have been still used.

b. Fully Electronic Under-Load Tap-Changer (or Solid-State Under-Load Tap-Changer or Static On-Load Tap-Changer)
There is no moving mechanical part in fully electronic tap-changers and whole tap-changer has been built by solid-state power switches. In Sects. 1.4 and 1.5, both groups are described and their drawbacks and advantages are given.

1.4 Electronically Assisted Under-Load Tap-Changers

One of the most important problems of mechanical under-load tap-changers is the arc in the contacts of diverter switches during the tap-changing process. The reason for appearance of arc is the mechanical nature of the switches. Of course, mechanical switches are interesting in the connecting of instant due to a very low voltage, however during tap-changing it has arc.

The main idea in the use of fully electronic under-load tap-changer is that during tap-changing process solid-state power switches with more controllability compared with the mechanical switches, come in and reduce the arc. But in a fixed tap, solid-state power switches exit the circuit and mechanical switches pass the load current. The reason is a very low connection voltage of mechanical switches compared with the solid-state power switches.

The first suggested circuit for the hybrid tap-changers is shown in Fig. 1.8. Figure 1.8 also presents the tap-changer diverter switches. The tap-changing switches in this structure do not differ from the under-load resistor tap-changer. In fact, Fig. 1.8 is the improved version of the under-load resistor tap-changer

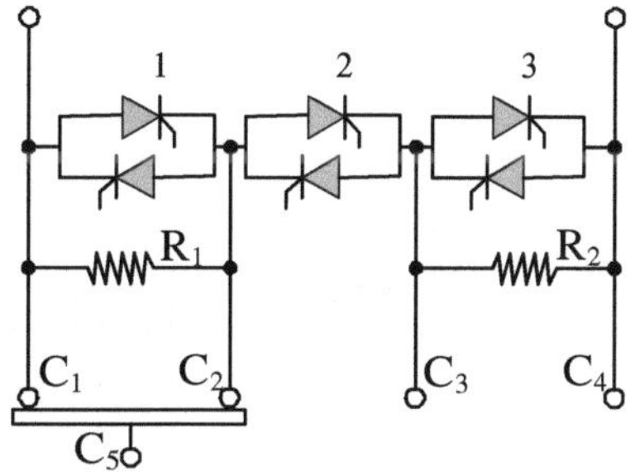

Fig. 1.8 First suggested circuit for hybrid tap-changers [17]

in Fig. 1.4b. In the structure shown in Fig. 1.8 (similar with the structure of Fig. 1.4b), in order to change the tap, the moving contact C_5 must move from left to right (or from right to left). However, in the structure of Fig. 1.8, before isolating C_5 from any of the fixed contacts, or connecting it to any of the fixed contacts, the relevant thyristors in Fig. 1.8 are turned-on and, therefore, no arc is caused. For example, at the moment of isolating contact C_5 from contact C_1, already a pair thyristors (1) has been turned-on and at the moment of isolating contact C_5 from C_1 by deviating the current from contact C_1 to a pair thyristors (1), the arc is minimized. The pair thyristors (1) is turned-off at the first crossing zero-current and the load current is transferred to resistor R_1. This happens before contact C_5 approaches contact C_3. Therefore, the time between isolating contact C_5 from C_1 and approaching contact C_3 is longer than an half cycle and this can already turned-off the pair thyristors (1). Pair thyristors (2) is turned-on before contact C_5 approaches C_2 and this does not allow an arc to be appeared as C_5 connects to C_3.

The structure suggested in Fig. 1.8 can largely reduce the arc, but its weakness is that although in this structure the pair thyristors are turned-on over a short time during tap-changing process, they are permanently connected to the diverting switches circuit and reduce the reliability of the system due to a possible burning of the thyristors. To solve this problem, the structure shown in Fig. 1.9 is recommended [18].

The main idea of this structure is that the pair thyristors are connected to the circuit only during tap-changing and in other instants it must be isolated from the circuit. In this case the tap-changer system does not fail if the thyristors burn and this increases the reliability of the system. To better explain the operation of the tap-changer of Fig. 1.9, first the contacts positions are described in the structure of Fig. 1.10 during switch-on and off periods.

Switch S between terminals 1 and 2 in Fig. 1.10 consists of fixed contact C_1 and main moving contact of C_2. In this structure, sensing contact C_3 and auxiliary contact C_4 have been included to reduce the arc. Auxiliary contact C_4 has been connected to terminal 2 through a pair thyristors and sensing contact C_3 has been connected to a pulse transformer primary Tr_1 terminal. Secondary of this transformer has been connected to the thyristors gate through a resistor and a diode. Contacts C_2, C_3 and C_4 are movable and connected mechanically to each others, but they are isolated electrically.

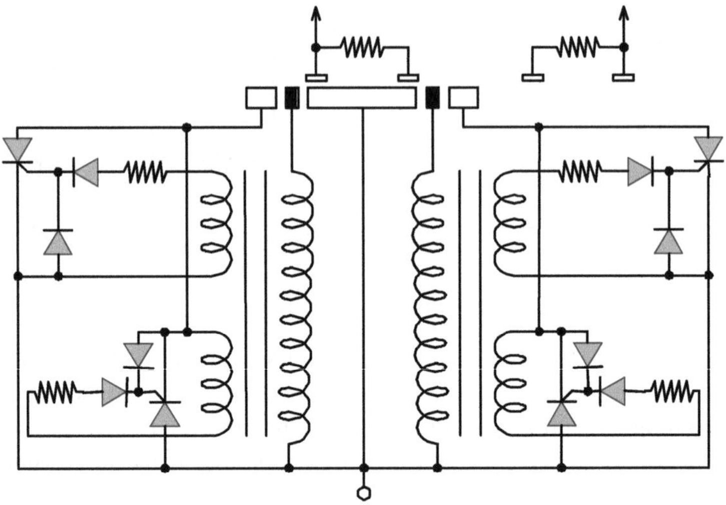

Fig. 1.9 Hybrid tap-changer structure [18]

Fig. 1.10 Contacts positions in suggested tap-changer of Fig. 1.9

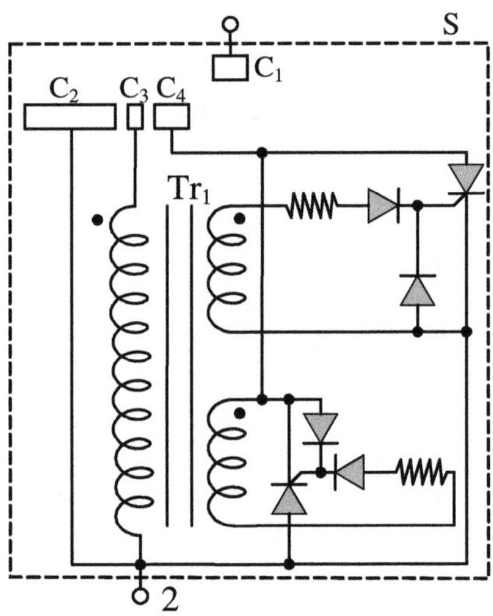

To connect switch S, contacts C_2, C_3 and C_4 move from left to right. Stages of connecting switch S are as follows:

a. Movable contacts C_2, C_3 and C_4 move to left and auxiliary contact C_4 connects to fixed contact C_1. Since in this case thyristors gates commands have not been issued, there is no current between terminals 1 and 2 of switch S and voltage difference between terminals 1 and 2 is applied to the pair thyristors.
b. By continuation of moving the movable contacts to the left, in addition to auxiliary contact C_1, sensing contact C_3 is also connected to fixed contact C_1. Therefore, two terminals of the primary of the transformer Tr_1 is short-circuited

and thyristors gates commands are disconnected. So the thyristors are switched-off the first zero load current crossing and the total current is transferred to contact C_2. After full transferring the current to contact C_2, contacts C_3 and C_4 are isolated from fixed contact C_1. Thus the process of connecting switch S is terminated.

Paying attention to the process of connecting switch S as described above, it is clear that at the instant of connecting or disconnecting movable contacts C_2, C_3 and C_4 from fixed contact C_1 in this process, these contacts do not carry current and therefore the arc reduces largely.

To disconnect switch S, movable contacts must move from right to left. Stages of disconnecting switch S are as follows:

a. By moving movable contacts C_3 and C_4 to the right, in addition to main contact C_2 that previously connected to fixed contact C_1, sensing and auxiliary contacts C_3 and C_4 are connected to contact C_1. However, primary of transformer Tr_1 has no voltage and pair thyristors in this case are off.
b. By continuation of moving the movable contacts to the left, there is an instant that main contact C_2 is isolating from fixed contact C_1. In this case contacts C_3 and C_4 connect to contact C_1. By isolating contact C_2 from contact C_1, there is a voltage difference between these two contacts that can generate arc. But before the extension of the arc, this voltage difference is applied to the primary of transformer Tr_1 and triggers the gate of one of pair thyristors, and the relevant thyristor is switched-on and transfers the current from contact C_1 to contact C_4. Therefore, the arc at the instant of isolating contact C_1 and C_2 is extinguished.
c. By continuation of moving the movable contacts to the right, sensing contact C_3 is isolated from fixed contact C_1 and thus the gate command of thyristors is switched-off and the thyristors will be off at the first zero current crossing and current of switch S is disconnected. After full disconnection of the current, contact C_1 is also isolated from contact C_1 and disconnecting process of switch S is ended.

Considering the above explanations about procedure for on and off of current of switch S in Fig. 1.10, operation of hybrid tap-changer Fig. 1.9 is clear. In this configuration, there are auxiliary and sensing contacts on two sides of the main moveable contact. While the main moveable contact moves to change tap from the left to right, auxiliary and sensing contacts at the left side of this main contact, improves the switching conditions in the switch-off instants of the main contact from the fixed contacts; auxiliary and sensing contacts at the right side of the main contact improves the switching conditions at the switch-on instants of the main contact from the fixed contacts. Therefore, the process of tap-changing in this tap-changer is done with a minimum arc.

Although in the suggested structure of Fig. 1.9 the arc is extinguished well and there will be no high maintenance costs, the tap-changing is still slow due to the use of a mechanical switched structure similar with an under-load resistor

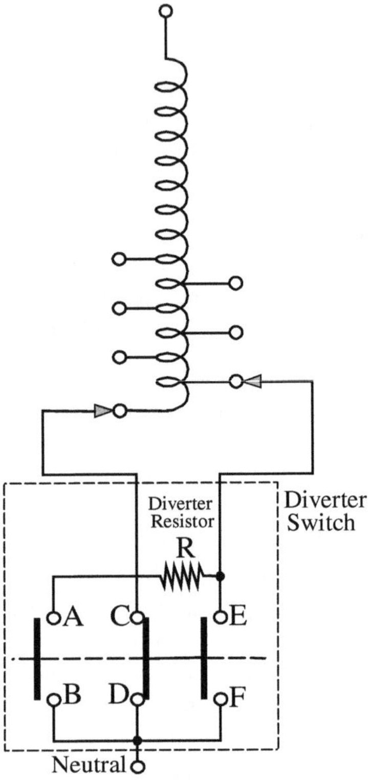

Fig. 1.11 Alternative hybrid tap-changer [19]

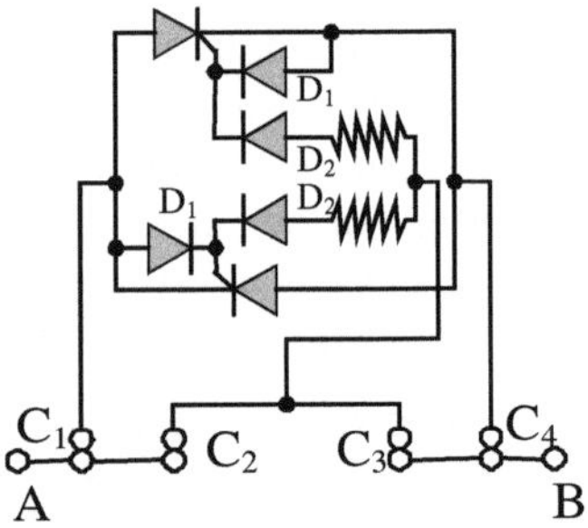

Fig. 1.12 Connecting pair-thyristor to the mechanical switches for terminal AB [19]

tap-changer (Fig. 1.4b). Meanwhile, the arrangement of the moveable contact in this configuration is complex.

The current disconnection in the switch in Fig. 1.12 is as follows:

An alternative structure, having different deviation switches in which only one transient resistor has been used, is shown in Fig. 1.11 [19]. An auxiliary pair-thyristor has been placed on each moveable contact which switch on terminals AB, CD and EF. Auxiliary pair-thyristor in Fig. 1.11 has not been presented in Fig. 1.12. Figure 1.12 shows the procedure for connecting pair-thyristor to the mechanical switches for terminal AB. The same is also true for other terminals. According to Fig. 1.12, contacts C_1, C_2, C_3 and C_4 are the mechanical contacts between terminals A and B. In this system, pair-contact (C_1, C_4) and also pair-contact (C_2, C_3) are always turned-on and off simultaneously. Contacts C_2 and C_3 are the main and permanent path of the current, while contacts C_1 and C_4 provide a current auxiliary path at the on and off instants through the pair-thyristor.

The current disconnection sequence in the switch of Fig. 1.12 is as follows:

a. In the steady-state, current passes contacts C_2 and C_3 as contacts C_1 and C_4 are open. To disconnect the current between terminals A and B, first contacts C_1 and C_4 are closed. There is no gate command in the pair-thyristor and thyristors are off due to zero voltage difference between contacts C_1 and C_2 and also contacts C_3 and C_4.
b. Contacts C_2 and C_3 start to open and at the same time, the voltage difference between these contacts and contacts C_1 and C_4 cause arc. Before that, this voltage difference triggers one of auxiliary pair-thyristor and the current transfers to contacts C_1 and C_4 and therefore the arc at contacts C_2 and C_3 extinguished. At the first zero passing, thyristor is switched-off and current is totally off, contacts C_1 and C_4 are also isolated. To ensure the zero current crossing and thyristor switching-off before opening contacts C_1 and C_4, the time between opening contacts C_1–C_3 and contacts C_1–C_4 must be at least a half-cycle.

Process of passing the current is as follows:

a. First contacts C_1 and C_4 are connected. However, the gate command of the auxiliary thyristors has not been yet issued, and these thyristors will be off and there will be no current.
b. In the next stage, contacts C_2 and C_3 start to close. Once an arc is generated in the mentioned contacts due to voltage difference, auxiliary thyristors gate command is issued and current passes through contacts C_1, C_4 and auxiliary thyristor, and, therefore, the arc is distinguished.
c. After full connection of contacts C_2 and C_3, there will be no gate command for the auxiliary thyristors and in the first crossing the zero, they move to the off mode and whole current will pass through contacts C_2 and C_3. At this point, contacts C_1 and C_4 will open and then the current will stop.

It is now clear how hybrid tap-changer in Fig. 1.11 operates. The stages of tap-changing are summarized in Table 1.2.

From Table 1.2 it is clear that in a moving stages both contacts CD and EF do not change simultaneously, therefore, a pair-thyristor can be used for each of these contacts. Such configuration has been shown in Fig. 1.13.

Configurations that have been so far suggested can extinguish the arc well, but the tap-changing will be still slow. The reason for this slow tap-changing is the use of conventional mechanical structures of the switches. To enhance the tap-changing speed, the configuration of Fig. 1.14 can be suggested [20], where selector switches and diverter switches have been improved. The selector switches are similar with that of the under-load resistor tap-changer from circuit arrangement point of view, however vacuum switch with a two-state control

Table 1.2 Stages of tap-changing in Fig. 1.11

	AB	CD	EF
(a) Passing left to right			
0		*	
1	*	*	
2	*		
3	*		*
(b) Passing right to left			
0	*		*
1	*		
2	*	*	
3		*	

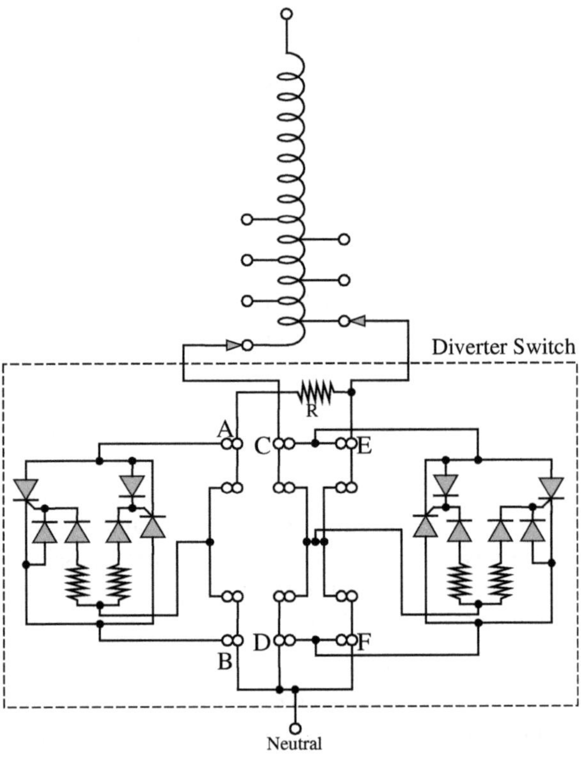

Fig. 1.13 A full structure of tap-changer with two auxiliary pair-thyristor [19]

electromechanical actuator is used in the place of contacts within oil and their complex mechanical drive.

Vacuum switches can be used in high voltage power systems. These switches have a long life. They have two windings and are turned-on and turned-off by excitation of the windings. The c hanging time of the vacuum switches from turn-on to turn-off or vice versa is about 20 ms.

Fig. 1.14 Configuration of a hybrid tap-changer [20]

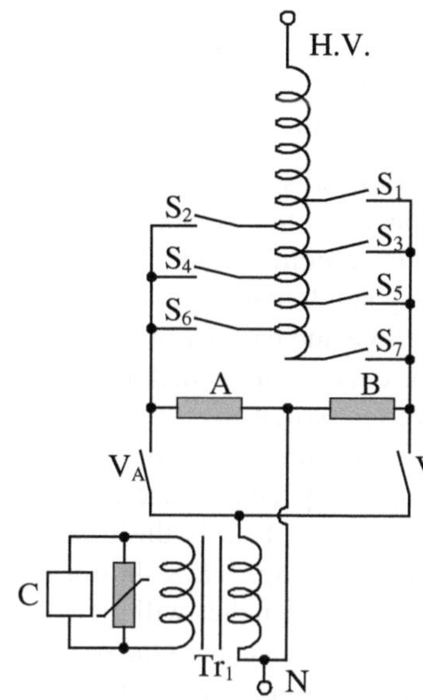

Diverter switches in the suggested configuration consist of two solid-state power switches A and B, and two vacuum switches V_A and V_B. To transfer the current from the vacuum switches to the solid-state switches, an auxiliary diverter switches circuit in series with vacuum switches V_A and V_B is used. The diverter circuit consists of a low power step-up transformer, solid-state power switch C and a varistor (for voltage drop). The solid-state power switches in this configuration are constructed using the switches that can be turned-off through the gate. The tap-changing stages in this configuration are as follows (as an examples the change from tap 2 to tap 3 stages is discussed):

a. When tap-changer is placed on tap 2, selector vacuum switch, deviation vacuum switch V_A and solid-state switches C are on and other switches are off. To change tap number 2 to tap number 3, first vacuum selector switch is closed, but no current will pass it.
b. In crossing zero current, solid-state switch A is on and solid-state switch C is turn-off. By turning-off switch C, the impedance of the current path through vacuum switch V_A increases and transfers the current to solid-state switch A. The transfer of the current from vacuum switch V_A to solid-state switch A takes a few ms. Following full transferring the current to solid-state switch A, vacuum switch V_A is off at zero current.
c. During crossing the next zero current, solid-state switch A is turn-off and at the same time solid-state switch B is turn-on. Following this, the current transfers to solid-state switch B and vacuum switch S_3.
d. Vacuum switch V_B is on, but the current passing through V_A and V_B has high impedance because solid-state switch C is turn-off, and there is no current in V_B.

e. During crossing the next zero current, solid-state switch B is switched-off and solid-state switch C is off simultaneously. Therefore, current is transferred to solid-state switch V_B. Finally selector switch S_2 is off and tap-changing stages is ended.

As discussed above, it is clear that in this configuration there is no crossing impedance and in fact during crossing zero current the current is transferred from the even number taps (left hand side number taps) to the odd taps (right side taps) using solid-state switches A and B. This is the reason that jump in tap-changing is possible and each even number tap can jump to any odd number tap and vice versa. The only limiting factor in taps jumping is the peak permissible voltage on the diverter switches.

In the proposed configuration in addition to extinguishing the arc, the tap-changing speed also rises. The required time for tap-changing from tap 1 to tap 19 is about 1.5 s. This is the case if the jumping possibility is not used. In the case, 3 jumps in taps takes shorter time, that is 0.5 s. Of course, in this case voltage of solid-state switches will increase. This is much shorter than 100 s of the under-load resistor tap-changer. One of the other advantages of configuration (Fig. 1.14) is that it gets rid of its complex mechanical system and replaces the vacuum switches. This enhances the mechanical reliability of the system. The reasons are a lower number of components in a vacuum switch and higher reliability. Of course, this configuration has some drawbacks. A higher cost due to the use of several vacuum switches and the permanent connection of the solid-state switches to the tap-changer and, therefore, low electrical reliability are the disadvantages of this system.

Another new configuration is shown in Fig. 1.15 [13]. In this configuration selector switches and diverter switches have not been separated. The fixed contacts have been designed such that they can pass the rated current through transformer. Diverter contacts each consists of three movable contacts. These contacts (one main contact and two auxiliary contacts) must also stand against the rated current. However, there is no need to turn the current on and off.

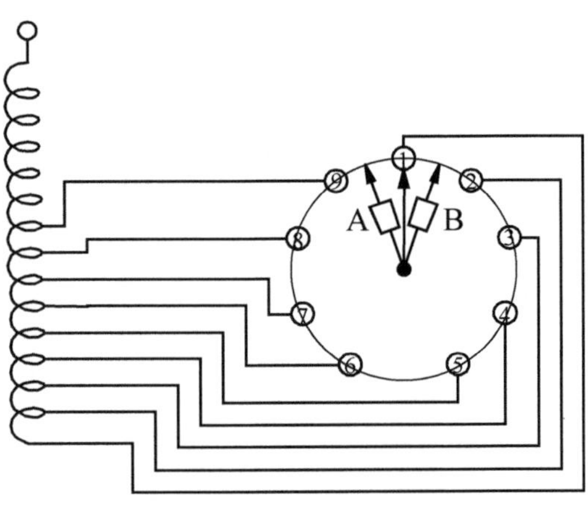

Fig. 1.15 Another hybrid tap-changer

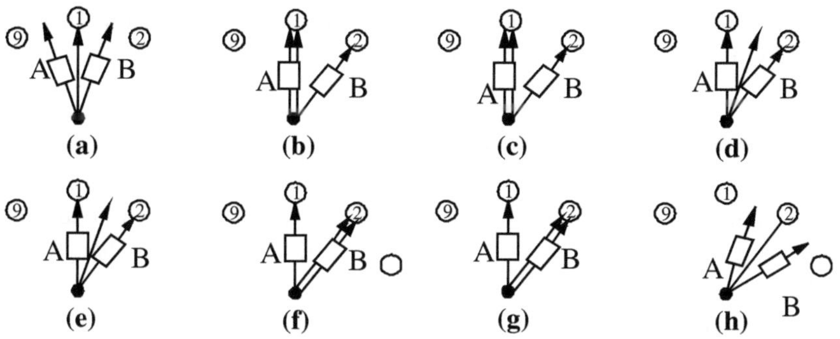

Fig. 1.16 Stages of tap-changing in hybrid tap-changer of Fig. 1.15 [13]

The movable contacts move circularly. A and B in Fig. 1.15 are solid-state switches which are similar to the configuration of Fig. 1.14, and are made up of switches which can be extinguished via the gate.

Stages of tap-changing in this configuration have been presented in Fig. 1.16 and described as follows (as an example the tap-changing process from tap 1 to tap 2 is described).

a. The main contact is connected to tap 1 and auxiliary contacts are placed on both sides of this contact. To change the tap to tap 2, the following stages are followed.
b. The auxiliary contacts move in CW, so that auxiliary is contacted A to contact 1 and auxiliary contact B to contact 2. In this case, solid-state switches A and B are off and no current passes the auxiliary contacts.
c. The main contact begins to move and at the same time solid-state switch A is on.
d. The main contact is disconnected from contact 1, but this is done with a minimum arc because solid-state switch A is on.
e. At the instant of zero current crossing, solid-state switch A is off and B is on. Therefore, the current is transferred from the auxiliary contact of switch A to the auxiliary contact of switch B.
f. The main contact continues its move approaching fixed contact number 3. In connecting the main contact to contact 2 no arc appears, because solid-state switch B is in the "on" position.
g. Solid-state switch B is off and total current is transferred to the main contact.
h. The main contacts move and are placed on both sides of contact 2. Thus the process of tap-changing from tap 1 to tap 2 ends.

Referring to the suggested circuits of the hybrid tap-changers, it can be concluded that the hybrid tap-changers can largely extinguished the arc during tap-changing, using solid-state power switches [21]. However, in these tap-changers there are still movable mechanical parts and this complicates the system and leads to the rather slow operation of the tap-changers. Meanwhile, some suggest to remove the passing resistors which decreases the tap-changing case losses.

Faster tap-changing is possible using the full-electronic version. In the next section this type of tap-changer is discussed.

1.5 Full-Electronic Tap-Changers

There is no movable part in full-electronic tap-changers and only solid-state power switches are used. The basic advantages of the full-electronic tap-changers are as follows:

a. Very Low Maintenance Cost
There is no movable mechanical part in full-electronic tap-changers, and no arc can appear during the tap-changing process as there is basically no contact; therefore the maintenance cost is very low (almost zero).

b. High Speed
The very fast switching process of solid-state power switches leads to the fast tap-changing in full-electronic tap-changers, as such that it is possible to change the tap at least once in any half-cycle.

c. Tap Jumping
There is no passing resistor in the full-electronic tap-changers and basically the circulation current between the taps is zero, so tap jumping becomes possible.

d. Better Performance
High speed and controllability of the solid-state switches and non-existence of mechanical limitations in the configuration of these switches enhance the capability and performance of the full-electronic tap-changers. Some of these capabilities are as follows:

1. Obtaining more steps with lower tap numbers and solid-state power switches. The reason is that there is non-limit in the configuration of the solid-state power switches.
2. A full-electronic tap-changer is a rapid static regulator as such that it can be considered as a custom power devices in power quality. It is capable to compensate the voltage sag, swell and also flicker.

e. Non-limit in tap-changing time.
The reason is that if power switches are correctly switched-on there will be slight fatigue in the switches. Of course, besides of the above-mentioned advantages, full-electronic tap-changers have some problems and limitations. These limitations are as follows:

a. Switch-on voltage drop of solid-state switch is larger than that of the mechanical switch, so operational losses of the full-electronic tap-changer is higher than that of the mechanical tap-changer.

b. Cost of full-electronic tap-changer is higher than that of the mechanical tap-changers because there are many solid-state power switch in the full-electronic tap-changer.
c. Full-electronic tap-changers must stand against short-circuit faults and large transient peaks in power system voltage due to the lightening.

Although the idea of the full-electronic tap-changers was proposed in 1973 [22, 23], they was restricted for special applications. The first comprehensive and academic study of this system was carried out in the 90 decay [24], then research continued. To introduce the progress and research trend on full-electronic tap-changers and to clarify its position, activities in this area are reviewed.

The conventional technique of using passing impedance for restricting the circulating current between the taps during tap-changing is also usable in the full-electronic tap-changers [24].

Figure 1.17 presents a typical configuration of this tap-changer. In this figure, all switches are AC solid-state power switches which are realized by pair-thyristor.

Bypass switch which may consist of a vacuum switch and/or a pair-thyristor with power higher than other thyristors, is used to bypass the fault conditions or transformer energization and protection of remaining thyristors. In tap-changer, Fig. 1.17, in order to change tap A to tap B, without taking into account the load power factor, the following stages are followed:

a. Initially the gate command of switch A is off and at the same time the gate command of switch X is on.
b. In the next cycle, the gate command of switch X and B are on both at the same time.

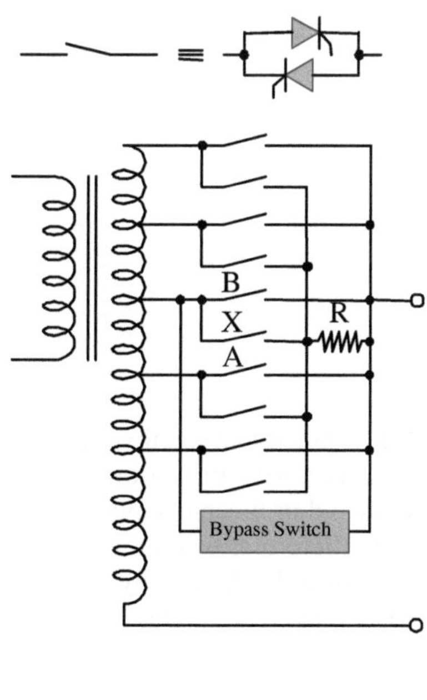

Fig. 1.17 The configuration of a typical full-electronic tap-changer using passing impedance [24]

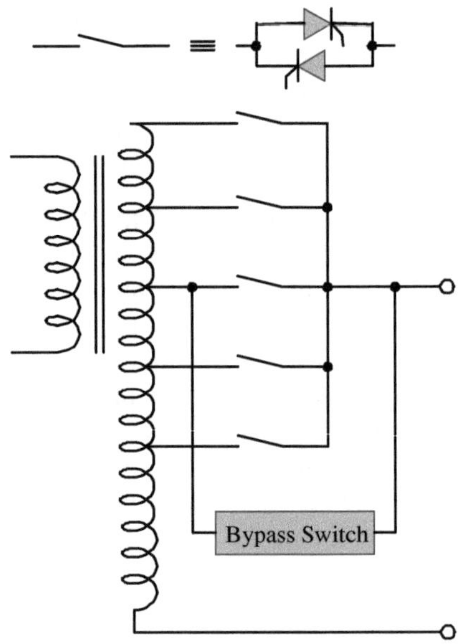

Fig. 1.18 An electronic tap-changer with no passing impedance [24]

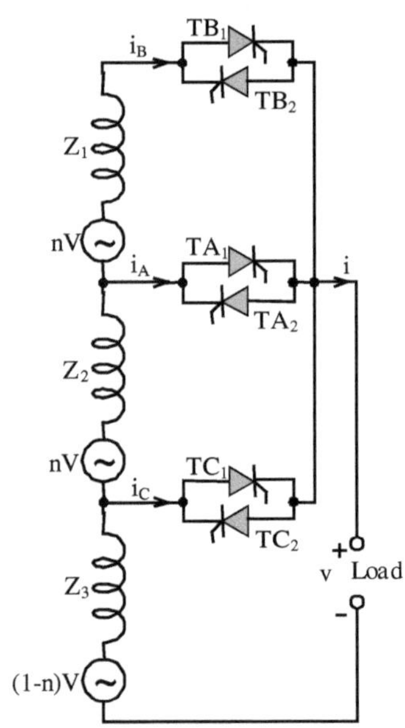

Fig. 1.19 Circuit model for secondary of tapped-transformer of Fig. 1.18

It is noted that the thyristors are turned-on by the gate commands, but they are not necessarily turned-off by the gate's off command. Basically the gate's off command or inverse applied voltage causes the thyristor current to approximate to zero. Therefore, when a gate command of pair-thyristor A is off and that of pair-thyristor X held, pair-thyristor X is not necessarily turned-off while pair-thyristor X is on and ready to let the current pass. Therefore, there is probably a circulating

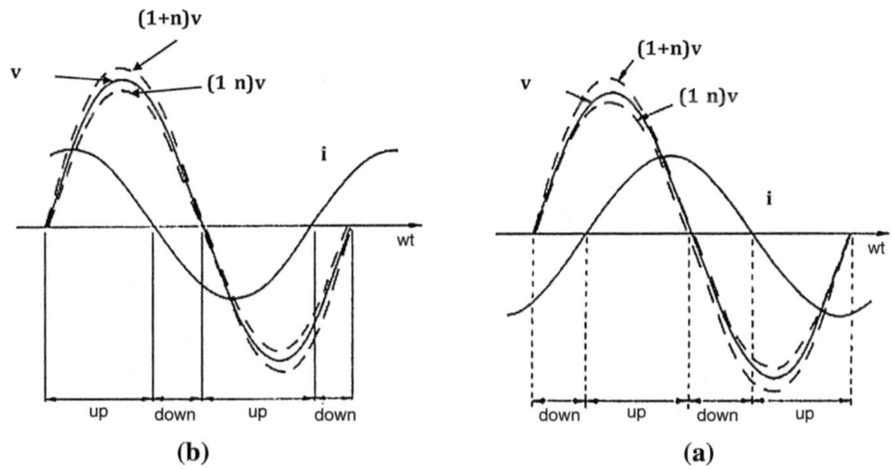

Fig. 1.20 Commutation of the current of switches during changing tap A to tap B in the presence of leakage impedance

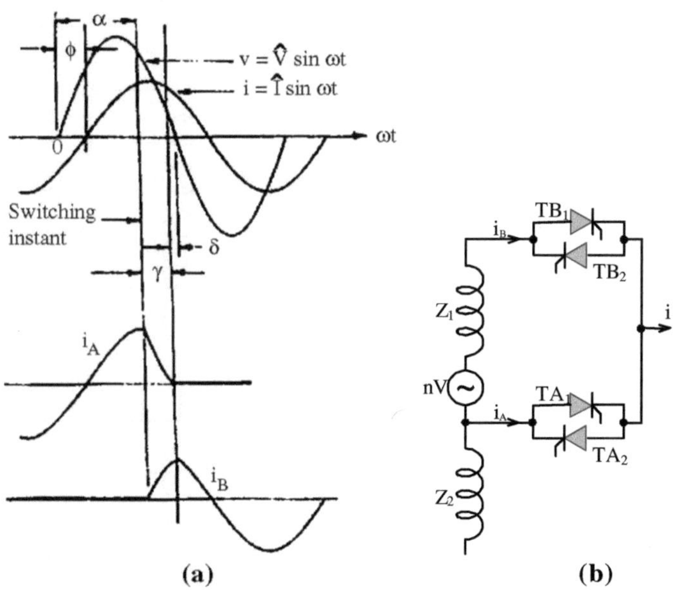

Fig. 1.21 Permissible switching time intervals for no circulating current: **a** inductive power factor and **b** capacitive power factor [24]

current between thyristors A and X limited by passing impedance R. After it can be insured that thyristor A is fully off and the second stage of tap-changing continues. It can be shown that if pair-thyristor X and B are on throughout zero-voltage crossing, there will be minimum stress on them [24]. However, a proper switching strategy for thyristors in configuration of Fig. 1.17 can be applied to the thyristors in Fig. 1.17 with no circulating current, where the passing impedance has been removed and the number of switches is reduced by half. Such configuration has been shown in Fig. 1.18. To evaluate the performance of this type of a tap-changer, the circuit mode of transformer is considered as Fig. 1.19. To simplify the figure, only three taps have been shown. The presented impedances are the leakage impedances of the secondary of transformer.

In Fig. 1.19, switch A (pair-thyristor TA_1 and TA_2) carries the load current, then the secondary voltage will go up or down rise by increasing or decreasing the tap (switch on of switch B) the tap (switch C is on). For tap-changing A to B or C, gate command of switch A stops and gate command of the proposed switch will be on. However, if no care is taken at this operation, there will be a probable taps short circuit for about one cycle. This short circuit current can damage the switches or the transformer winding. To prevent this short circuit, the switching instant must be precisely controlled. If the leakage impedances in model of Fig. 1.19 are

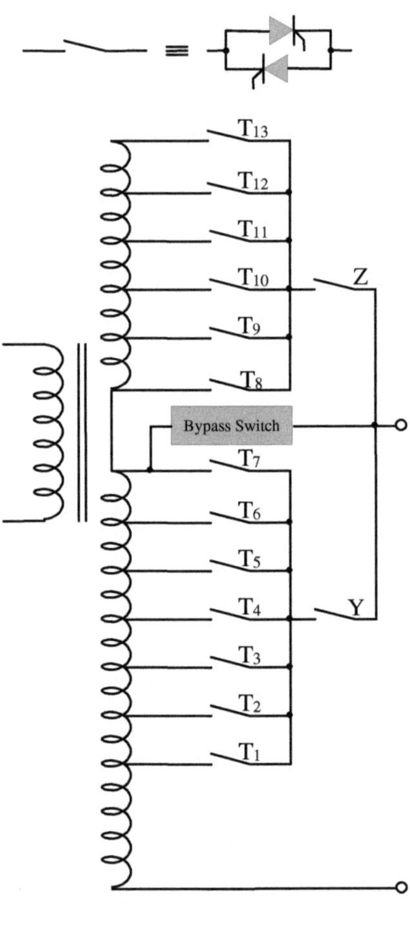

Fig. 1.22 A full-electronic tap-changer configuration for reduction of switches voltage [24]

ignored, considering the switching off of thyristors, the permissible switching time intervals for no circulating current versus the load power factor can be shown as Fig. 1.20.

Referring to Fig. 1.20, the switching logic can be simply expressed as follows:

1. If voltage and current have the same sign, the permissible intervals are for increasing the tap.
2. If voltage and current have the opposite sign, the permissible intervals are for decreasing the tap.

In the above switching logic, if the power factor is close to unity, in the case of direct power flow, the permissible time for reduction of tap tends to zero, and in the inverse power flow, the permissible time for increasing tap tends to zero. In such cases, the short circuit current is somehow unavoidable.

Of course in the actual conditions, there is leakage impedance of transformer, and current commutation between the switches will not occur instantaneously and there is a delay. This commutation time depends on the leakage impedance, taps voltage and load current. In this case, the permissible times for tap-changing must

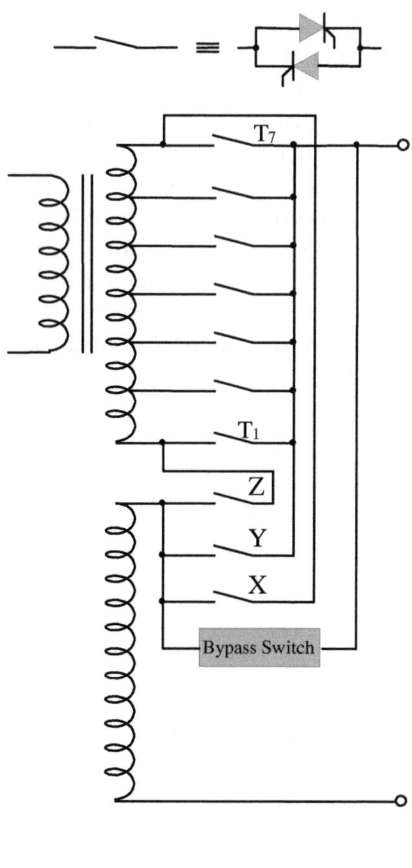

Fig. 1.23 A full-electronic tap-changer configuration for reduction of the number of switches and taps [24]

be determined by taking into account the commutation time. Figure 1.21 shows the changing tap A to B by taking into account the commutation time.

The commutation time (overlap angle) in this figure has been represented by γ. It is noted that in order to protect the over-current in the thyristors energizing transformer and over-current due to the secondary short-circuit fault, the bypass switch in Fig. 1.18 has the same explanation as that of Fig. 1.17. This switch is with snubber depending on the tap-changer voltage and current, or a vacuum switch, or a pair-thyristor with the rating higher than the remaining thyristors.

The main configuration of a full-electronic tap-changer has been shown in Fig. 1.18. Its switching strategy is as such that its tapes are correctly and with a minimum circulating current changed. This configuration is an initial and simple arrangement compared to the solid-state power switches. It means that there are voltage steps and AC solid-state power switches equal to the transformer taps number. This will cost too much when high number of steps is required because the number of the required thyristors and their maximum voltage levels is higher.

To solve this problem, two alternative configurations including more advanced arrangements are suggested as shown in Figs. 1.22 and 1.23.

The basis of switching strategy for proper commutation of the switches in the above-mentioned configurations is exactly similar to those expressed in Fig. 1.18. However, as seen the switches arrangements in these two configurations are more complicated than that in the initial configuration of Fig. 1.18.

In Fig. 1.22 configuration, switches have been classified into two groups and groups connected to the load were chosen by switches Y and Z; for example, for changing 7th tap to 8th tap, switches Tv and Y are off, and switches TA and Z are on (in order to properly on and off the switches without short-circuit and circulating current, this must be done by a proper switching logic).

In this configuration, there is 13 voltage steps, but it is enough that each group switches stand a portion of this voltage. Thyristors group T_1–T_7 must stand six times of every step voltage and thyristors group T_8–T_{13} must stand five times of every step voltage. Therefore, in total thyristors voltage is reduced in this configuration. In spite of this advantage, this configuration has also two drawbacks:

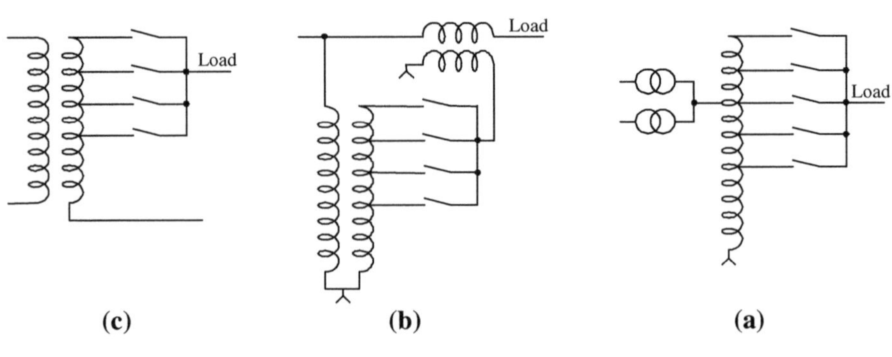

Fig. 1.24 Three different cases for placing electronic tap-changer as a fast voltage regulator

1. Always two switches are in series, and this causes the rise of the voltage drop. Therefore, it increases the tap-changer losses.
2. Voltages of thyristors Z and Y are high and equal to 12 steps. Of course, by improving the on and off strategy of the thyristors, this voltage drop can be reduced. At this end, when switch Y is off, switch T_7 is kept on in order to supply a minimum voltage on switch Y. Similarly, when switch Z is off, switch T_8 must be on.

However, the idea used in configuration of Fig. 1.22 can be extended as such that the thyristors are divided into two groups and then different groups are connected to the load by the switches.

The idea employed in the configuration of Fig. 1.23 is that by providing the possibility of inverting the taps polarities, around $2n$ voltage steps are achieved with n taps and switches. When switches T_7 and X are on (it means voltage is at the middle), and there is intension of inverting the taps, first switch Y is on and then switches X and T_7 are off. Then, T_1 and Z are on and finally Y will be off. This configuration is also similar to the configuration of Fig. 1.23 and has this drawback that it is always in the series path with the current of two thyristors and this increases the losses. Another drawback of this configuration is that when switch X is on, the load voltage will be lower than the middle voltage, while the winding drop is larger compared with the case in which switch Y is on and the middle voltage is connected to the output due to the winding of the taps.

Electronic tap-changer can be proposed as a fast regulator for protection of the sensitive loads against voltage rise and drop. There are three different cases for placing electronic tap-changer as shown in Fig. 1.24 [25].

If a proposed sensitive load (for example solid-state devices of a factory) already has one or more full transformers, electronic tap-changer can be fixed on a transformer in series with the existing transformers (Fig. 1.24a). If the proposed

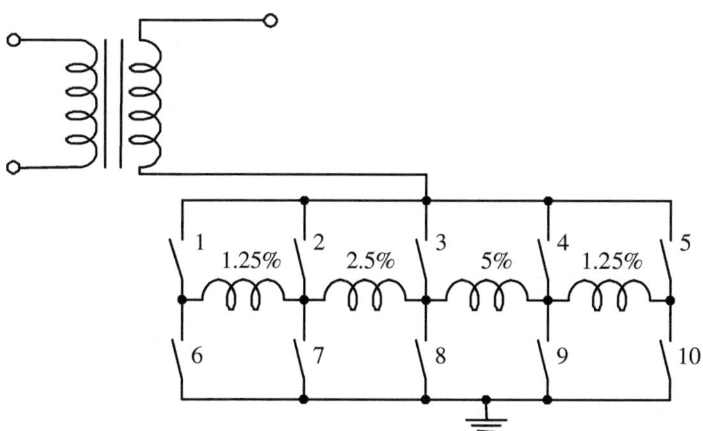

Fig. 1.25 Full-electronic tap-changer configuration suggested by [26]

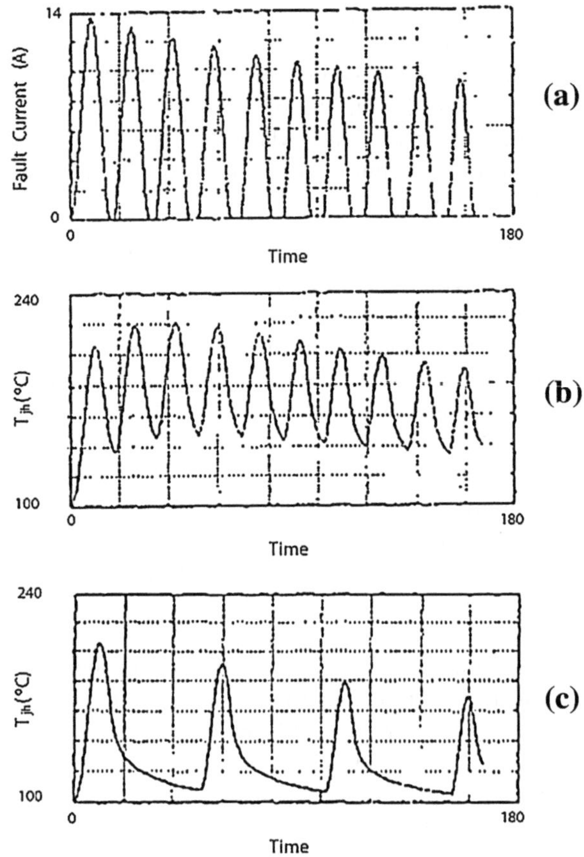

Fig. 1.26 **a** Short circuit current, **b** contact temperature up to heat sink when current continuously passes a pair-switch, **c** contact temperature up to heat sink when current alternatively and cycle by cycle passes a pair-switches [26]

voltage and current are not equal to that of the tap-changer, a booster transformer can be used. This configuration has been presented in Fig. 1.24b.

Finally, in the cases that the voltage variation is required, electronic tap-changer is placed on a full transformer (Fig. 1.24b). There is a new idea for cost reduction

Table 1.3 Economical comparison of mechanical and full-electronic tap-changers of Fig. 1.15 for a 10 MVA, 27/3.8 kV system

No.	Title	Cost of mechanical tap-changer (k$)	Cost of the suggested tap-changer (k$)
1	Transformer and tap-changer	285	312
2	On-load losses of transformer	45.5	45.4
3	No-load losses of transformer	31.5	31.5
4	Thyristors losses	0	3.5
5	Total cost	361	391.4

of the full-electronic tap-changers. In this idea the solid-state power switches arrangement have been improved, and also a new modulation method for improvement of the control strategy are introduced [26]. Figure 1.25 exhibits the configuration of this electronic tap-changer.

In this configuration, it is possible to create 16 steps in the range of −10 to +10% of the rated voltage. The voltage of each step is 1.25% of the rated voltage. Therefore, the first advantage of this tap-changer is that only by using 5 taps in the transformer and 10 solid-state AC power switches, 16 voltage steps have been obtained. Another advantage of this configuration appears in short-circuit conditions. In the output short circuit conditions, each of parallel switches pair can pass the short circuit current; therefore, in short circuit case the current can be transferred to each of these pair switches cycle by cycle, so the high stress does not exert on one of the switches pair. Figure 1.26b and c show the contacting temperature up to thyristors heat sink for two cases: (I) passing short circuit current

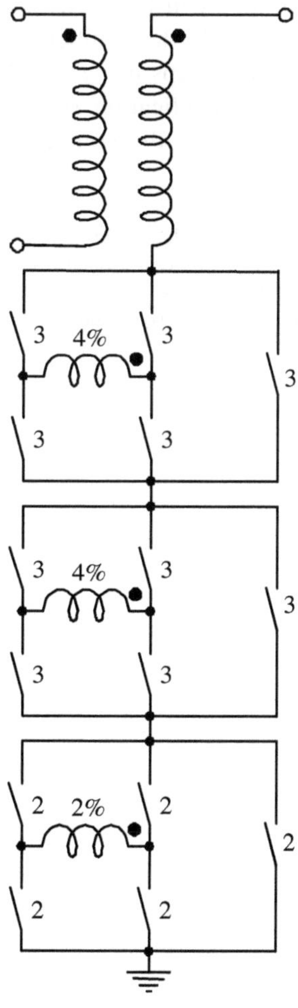

Fig. 1.27 Electronic tap-changer configuration for a 115 kV, 100 MVA with maximum ± 10 changes by tap [27]

from one of the switches pair and (II) passing short circuit current alternatively between the switches pair. According to Fig. 1.26, the maximum temperature for case I is 222°C, while this temperature is 204°C for case II. Since the thyristors current in a full-electronic tap-changer is chosen based on the current in the short circuit conditions, the rated current of thyristors in this new configuration can be reduced. The other advantage is clear by referring to Fig. 1.26c. After three cycles in case II, temperature of the contact approaches the environment temperature, and it rises again. It means that if the short circuit condition continues, there will not be any problem for thyristors because their contact temperatures have no rising trend, and it is up and down periodically [27].

A new modulation method for improving control strategy and increasing voltage steps without changing the tap-changer configuration has been introduced in literatures [26]. The basis of this modulation method called discrete cycle modulation (DCM) is that the proposed rms voltage in the output is formed by adding the number of cycles of a tap voltage and other tap voltages, therefore, rms voltage will be the average between rms voltage of each individual tap. Without increasing the number of switches and taps, it is practically possible to build more voltage steps using DCM.

By application of different techniques for cost reduction of electronic tap-changer in the configuration of Fig. 1.25, the cost difference between this tap-

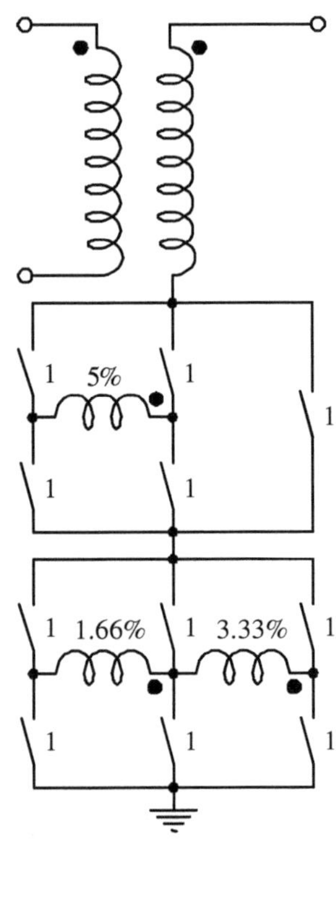

Fig. 1.28 Electronic tap-changer configuration for a 34.5 kV, 30 MVA with maximum ± 10 changes by tap [27]

changer and similar mechanical under-load tap-changer is largely decreased. Table 1.3 compares the cost of these two types of tap-changer for a 13.6 kV, 10 MVA system.

Table 1.3 shows that the relative difference of the costs between the mechanical tap-changer and full-electronic tap-changer is only 8%. If a very low maintenance cost of the full-electronic tap-changer and its large capabilities are taken into account, the popularity of the full-electronic tap-changer will be fully realized. This shows that by improving the full-electronic tap-changer configuration, the cost difference between the full-electronic and mechanical tap-changers will be considerably reduced.

Figures 1.27 and 1.28 show two new configurations for a full-electronic tap-changer. The main goal in the design of these configurations is the cost reduction of the tap-changer by decreasing the number of switches and their voltage and current and also reduction of the number of transformer taps. Figure 1.27 presents the suggested configuration for a 115 kV, 100 MVA system, and Fig. 1.28 shows the recommended configuration for a 34.5 kV and 30 MVA system. Each of the presented switches in this figure is formed by a number of thyristors pair connected anti-parallel. The number of thyristors pair in series has been mentioned beside each switch. In these two configurations, the thyristors with standing voltage of 450 V are used. Configurations in Figs. 1.27 and 1.28 come from the basic configuration of

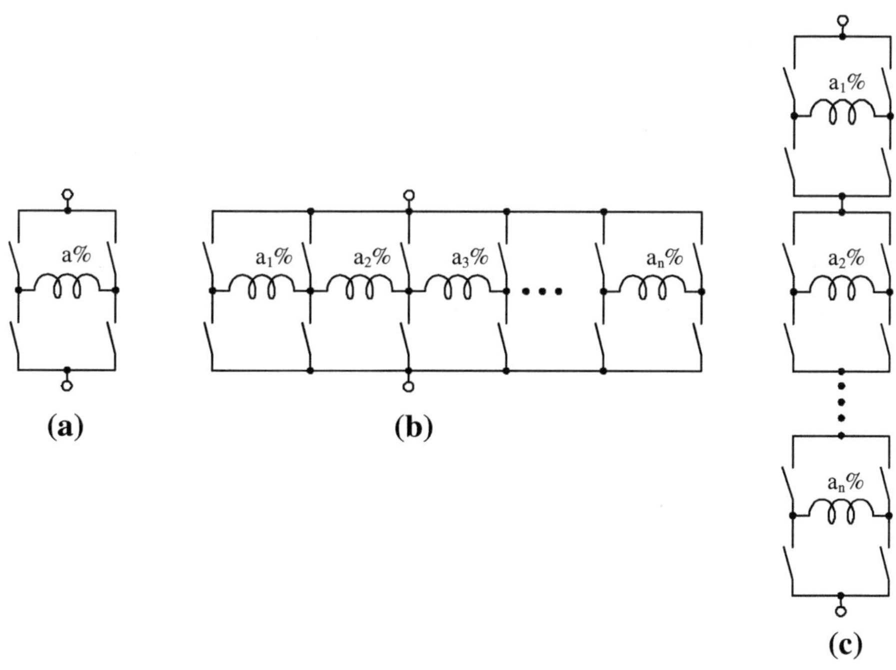

Fig. 1.29 **a** Basic configuration of electronic tap-changer used in Figs. 1.27 and 1.28, **b** basic configuration of extended electronic tap-changer in parallel, and **c** basic configuration of extended electronic tap-changer in series

Fig. 1.29a. The basic configuration consists of a winding with a percentage of the rated voltage and four solid-state AC switches. Figure 1.29a can be extended as parallel (Fig. 1.29b) and series (Fig. 1.29c). The requirements of each application will determine how tap-changer must be built using this basic configuration. A DCM similar to the above-mentioned DCM has been proposed [27]. It will be described in detail and its advantages disadvantages are noted.

The main idea in DCM method is that in order to increase the number of voltage taps in a full-electronic tap-changer, without increasing the number of switches and taps of the transformer, a proper switching of the tap-changer switches can be applied and the output voltage is regulated as such that a number of cycles on one tap and number of other cycles on the other tap are placed and this trend continues periodically. So, rms voltages between each of the mentioned taps are obtained in average as given below. In this calculation, the peak voltage of the lower tap is presented by V_0 and higher tap with V_1. Meanwhile, N is the total number in a DCM period and M is the number of the repeated cycles of the higher tap. Figure 1.30 shows a typical DCM waveform in which $V_1 = 1$, $V_1 = 1.02$, $N = 3$ and $M = 2$.

When DCM is used, output voltage of transformer $v(\omega t)$ over a modulation period is represented as follows:

$$V(\omega t) = [V_1 - au(\omega t - 2\pi M)u(2\pi N - \omega t)]\sin(\omega t) \tag{1.1}$$

where $a = V_1 - V_0$ and $u(t)$ is a unit step function. By evaluation of rms voltage $v(\omega t)$ which repeats by period NT, the following result is obtained:

$$V_{\text{rms}} = \sqrt{\frac{MV_1^2 + (N - M)V_0^2}{2N}} \tag{1.2}$$

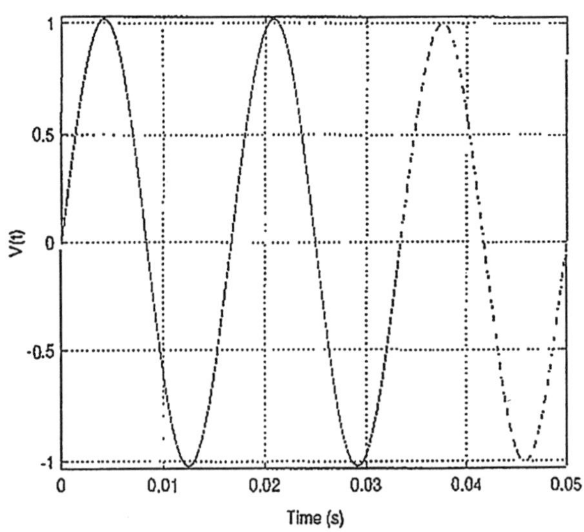

Fig. 1.30 A period of DCM waveform for $V_0 = 1$, $V_1 = 1.02$, $N = 3$ and $M = 2$

Equation 1.2 indicates that the rms voltage is a function of V_0, V_1, N and M, so for constant values of V_0 and V_1 and varying M or N, the rms voltages between $V_0/\sqrt{2}$ and $V_1/\sqrt{2}$ are obtained. It means that there are new voltage steps. In spite of the merit of DCM method in rising the number of steps, this method has three limitations.

1. DCM waveform contains sub-harmonics of the fundamental frequency. These sub-harmonics can be expressed by Fourier series expansion. The Fourier series expansion is presented as follows:

$$v(\omega t) = b_N \sin(\omega t) + \sum_{\substack{n=1 \\ n \neq N}}^{\infty} C_n \sin\left(\frac{n}{N}\omega t + \phi_n\right) \tag{1.3}$$

It can be shown that coefficients b_N, C_n and ϕ_n are obtained from Eqs. 1.4–1.6.

$$b_N = \frac{MV_1 + (N-M)V_0}{N} \tag{1.4}$$

$$C_n = \left|\frac{N}{(N^2 - n^2)n} - (V_0 - V_1)\right| \sqrt{2 - 2\cos\left(2\pi M \frac{n}{N}\right)} \tag{1.5}$$

$$\phi_n = \mathrm{Arctg}\left(\frac{\cos\left(2\pi M \frac{n}{N}\right) - 1}{\sin\left(2\pi M \frac{n}{N}\right)}\right) \tag{1.6}$$

Values of coefficients b_N and C_n up to $n = 10$ for three following cases:

$$\left\{\begin{matrix} N=3 \\ M=1 \end{matrix}\right\}, \left\{\begin{matrix} N=3 \\ M-2 \end{matrix}\right\}, \left\{\begin{matrix} N=2 \\ M=1 \end{matrix}\right\}$$

have been given in Table 1.4. This table has been obtained for $V_0 = 1$, $V_1 = 1.02$ or a $= 2\%$ of the voltage. By referring to Table 1.4, it is seen that the largest value

Table 1.4 Harmonics coefficients for DCM waveform in three cases

No.	$M = 1, N = 3$	$M = 2, N = 3$	$M = 1, N = 2$
1	0.004134	0.004234	0.008499
2	0.006615	0.006615	1.010000
3	1.006666	1.013333	0.005092
4	0.004725	0.004725	0
5	0.002067	0.002067	0.001212
6	0	0	0
7	0.000826	0.000826	0.000565
8	0.000601	0.000601	0
9	0	0	0.000330
10	0.000363	0.000363	0

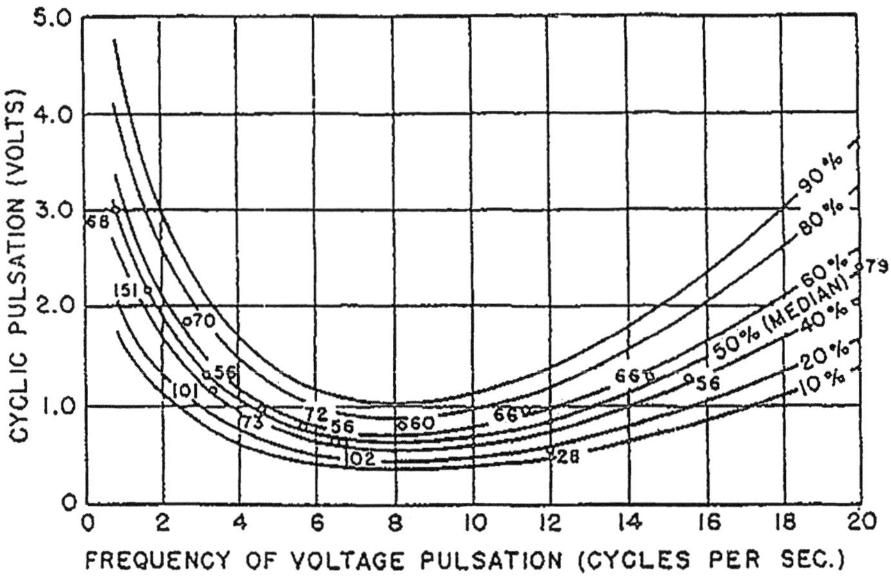

Fig. 1.31 Results of flicker sensing test for voltage variations of an incandescent lamp [27]

is obtained for $\left\{ \begin{matrix} N = 2 \\ M = 1 \end{matrix} \right\}$ and, an half of the fundamental frequency, and it is 0.008488.

2. In DCM waveform, there is a periodic variation of the voltage amplitude and this causes flicker in incandescent lamps. Of course, flicker depends upon several factors such as human factors, voltage variations, and lamp power and its permissible range has not been fully defined. Figure 1.31 shows the test results of 1,104 observations by 95 persons. These results are obtained by applying the variations of a pulse amplitude voltage for different amplitudes and frequencies on 1,157 incandescent lamps with powers 25, 40 and 60 W. The % inserted beside each curve shows % of the persons who have sensed the flicker. As seen, voltage variations less than (0.29%)/3 V is sensed by less than 10%. It is clear that DCM waveform must be as such that its flicker places in the permissible range.
3. Time response of a system by applying DCM must be large enough in order to practically realize the rms mean voltage over N periods of the power and the value of each tap is filtered individually.

DCM method has not been used any more due to the three above-mentioned drawbacks.

To enhance the power quality, the design of full-electronic tap-changer for distribution transformers can be initialized [28–33]. Solid-state switches in electronic tap-changer are switched by high frequency (kHz); therefore, the mentioned tap-changer is an high frequency AC/AC converter. IGBT can be used as solid-state switch. High frequency switching in electronic tap-changer enhances the capabilities of the full-electronic tap-changer. Capability of improving the waveform (elimination of harmonics and flicker and ...) and regulating and controlling the output voltage continuously (not discreted) are its additional capabilities. On the other hand, more system complexity, generating high frequencies harmonics and rising the tap-changer losses (arising from switching losses) are the drawbacks of these types of tap-changers. These put the limitations on its applications. Of course, to decrease the stress on the switches, soft switching idea for high frequency tap-changer can be recommended [34–37]. In this case, by adding a resonance circuit and also additional solid-state switches, it is possible to provide zero current switching (ZCS) conditions for main solid-state switches and this decreases the stress on the switches and switching losses. The system is considerably complicated and its application is restricted to a special cases.

The idea of electronic tap-changer has been combined with AC/AC high frequency converters concept and an AC voltage regulator for reducing and increasing the voltage introduced [38]. However, there is no new idea on the design of the electronic tap-changer. Electronic tap-changer has been introduced for village distribution lines with weak cables [39], but there is no a new suggestion for the design of the electronic tap-changer. The simple and initial configuration of Fig. 1.17 has been used and the major point is on the fixing the electronic tap-changer [40]. An electronic, as a fast regulator, has been discussed in literatures [41–44].

1.6 The Goal of this Book

Although various ideas have been introduced for the realization of electronic tap-changer, there is no versatility in these ideas. It is clear that which ideas are convenient for a specific application. Basically, which configurations are optimal? or do the optimal configurations have different structures?

The purpose of an optimal configuration is the structure that has the advantages of full-electronic tap-changers and drawbacks of these tap-changers are decreased as possible. One of the most important problems of electronic tap-changer is its cost.

In this book ideas introduced for electronic tap-changer are first classified logically. Then, cost parameter as objective function is defined to present an optimal configuration. Following the selection of the optimal configuration, remaining design requirements such as switching method for a proper commutation, control method and detection of the rise and drop of voltage, etc. are designed tap-changer and finally the designed is practically realized.

1.7 Future Chapters

The next chapter deals with the design of the structure of winding taps and also switches configuration in the full-electronic tap-changer. Cost function is defined and optimal configuration based on this function is designed. Chapter 3 introduces bi-directional switches, and current commutation techniques between them are discussed. Chapter 4 presents the design of control system of tap-changer for the regulation of the output voltage. In this chapter, tap-changer system is simulated and the performance of the designed control system is investigated. Meanwhile, it is discussed that how amplitude and phase of a sinusoidal variable are extracted. Chapter 5 proposes an electronic tap-changer as power quality enhancement tools and its performance is determined. Chapter 6 presents the experimental results on a prototype tap-changer and finally the book is concluded in Chap. 7.

References

1. IEEE Standard Dictionary of Electrical and Electronics Terms, 3rd edn (1984) IEEE Press, New York
2. Stigant A, Franklin AC (1998) J & P transformer book, 12th edn. Elsevier, Butterworth
3. Feinberg R (1979) Modern power transformer practice. Halsted Press, New York
4. Gebert KL, Edwards KR (1974) Transformers principles and applications, 2nd edn. American Technical Publishers Inc., Illinois
5. Central Station Engineers of the Westinghouse Electric Corporation (1964) Electrical transmission and distribution reference book, 4th edn
6. Bimbhra PS (1992) Generalized theory of electrical machines. Khanna Publishers, New-Delhi, India
7. Alstgom, The transformer guide
8. Say MG (1984) Alternating current machines, 5th edn. Pitman, UK
9. Laughton MA, Say MG (2003) Electrical engineer's reference book, 16th edn. Elsevier Science
10. Fink DG, Beaty HW (2004) Standard handbook for electrical engineers, 14th edn. McGraw-Hill
11. ABB (1997) On-load tap-changers, Type UBB-Technical guide (1997)
12. Kang P, Birtwhistle, D (2003) Condition assessment of power transformer on-load tap-changers. IEEE Trans Power Deliv 18(1):78–8
13. Gao D, Lu Q, Luo JA (2002) New Scheme for on-load tap-changer of transformers, international Conference on Power System Technology. Power Conference Publications, pp 1016–1020
14. Jiang H, Shuttleworth R, Al Zahawi BAT, Tian X, Power A (2001) Fast response GTO assisted novel tap-changer. IEEE Trans Power Deliv 16(1):111–115
15. Shuttleworth R, Tian X, Fan C, Power A (1996) New tap changing scheme. IEE Proc Electric Power Appl 143(1):108–112
16. Harlow J (2001) Discussion of fast response GTO assisted novel tap changer. IEEE Trans Power Deliv 16(4):826–827
17. Robert ME, Ashman WG (1969) A thyristor assisted mechanical on-load tap-changer. IEE Proc Power Thyristors Their Appl 185–192
18. Cooke GH, Williams KT (1990) Thyristor assisted on-load tap-changers for transformers, 4th international conference on power electronics and variable-speed drives, 17–19 July 1990, pp 127–131
19. Cooke GH, Williams KT (1992) New thyristor assisted diverter switch for on-load transformer tap-changers. IEE Proc Electric Power Appl 139(6):507–511

20. Shuttleworth R, Power A, Tian X, Jiang H, Talzahawi BA (1997) A novel thyristor-assisted tap-changer scheme, IEE 14th international conference on electricity distribution, June 1997, vol 1, pp 28/1–28/5
21. Faiz J, Javadnia H (2000) Fast response solid-state on-load transformers tap-changer, 8th IEE international conference on power electronics and variable speed driver, pp 355–359
22. Musgrave G, O'Kelly D (1974) Improvement of power system transmission by solid-state techniques. IEE Conf Publ 123:228–233
23. O'Kelly D, Musgrave G (1973) Improvement of power system transient stability limits. Proc IEE 129(2):247–253
24. Zai FQY, O'Kelly D (1996) Solid-State on-load transformer tap-changer. IEE Proc Electric Power Appl 143(6):481–491
25. Larsson T, Innanen R, Norstrom, G (1997) Static electronic tap-changer for fast phase and voltage control. IEEE electric machines and drives conference, 18–21 May 1997, pp TC3/4.1-TC3/4.3
26. Degeneff RC (1997) A new concept for a solid-state on-load tap-changer. IEE Conf Publ 1(438):7/1–7/4
27. Dernirci O, Torrey DA, Degeneff RC, Schaeffer FK, Frazer R (1998) A new approach to solid-state on-load tap-changing transformers. IEEE Trans Power Deliv 13(3):952–961
28. Bauer P, deHaan SWH (1998) Electronic tap changer for 500 kVA/10 kV distribution transformers: design, experimental results and impact in distribution network. IEEE IAS 2:1530–1537
29. Bauer P, deHaan SWH, Paap GC(1996) Solid-state control for transformer tap changing. PEMC 1996, Budapest, Sept 1996, pp 39–44
30. Bauer P, deHaan SWH (1997) Electronic tap- changer for 10 kV distribution transformers, EPE 97, Sept 1997, pp 3/1010–3/1015
31. Bauer P, de Haan SWH (1997) Protective device for electronic tap-changer for distribution transformers, EPE 97, Sept 1997, pp 4/282–4/287
32. Bauer P, deHaan SWH (1999) Solid-state tap-changers for utility transformers. IEEE AFRICON 2:897–902
33. Bauer P, Schoevaars R (2003) Bidirectional switch for a solid-state tap-changer. IEEE PESC 03, pp 466–471
34. Villegas Vaquero GJ, Echavarria R, Horta S, Perez MA, Martinez S (1998) Quasi-resonant fast on-load two tap changing stabilizer. IEEE Int Cong Power Electron 177–183
35. Sanchez V, Echavarria R, Cotoragea N, Claudio A (2000) Design and implementation of a fast on-load tap-changing regulator using soft-switching commutation techniques. PESC 2000, June 2000, pp 488–493
36. Echavarria R, Sanchez V, Ponce M, Cotorogea M, Claudio A (2004) Analysis and design of a quasi-resonant fast on-load tap changing regulator. J Circuit Syst Comput 13(4):877–899
37. Echavarria R, Sanchez V, Ponce M, Claudio A, Cotorogea M (2002) Parametric analysis of a quasi-resonant fast on-load tap-changing regulator. PESC 2002, pp 1809
38. Jang DH, Choe GH (1998) Step-up/down AC voltage regulator using transformer with tap changer and PWM AC chopper. IEEE Trans Industrial Electron 45(6):905–911
39. Meyer AS, Coller JV (1999) Electronic tap changers for use with ultra-light rural distribution lines. IEEE AFRICON 1999, Oct 1999, vol 2, pp 909–912
40. Xiaoming L, Qingfen L, Xianggen Y, Jianghui X (2002) A New on-load tap-changing system with power electronic elements for power transformers. IEEE Int Conf Power Syst Technol 1:556–559
41. Quaia S, Tosato F (1996) Unsymmetrical three-phase voltage sag compensation through no energy stored voltage boosters, 84th mediterranean electrotechnical conference of MELECON 96, May 1996, Italy, vol 2, pp 921–926
42. Alegria E, Khan A, Rajda J, Dewan S (1998) Static voltage regulator (SVR)-ride through support for semiconductor facilities, power systems world' 98 power quality conference of Santa Clara, Nov 1998, pp 1–7

43. Toh CW, Lock CS, Chen S, Choi SS (2000) A voltage quality improvement solution using step dynamic voltage regulator, IEEE power conference 2000, Dec 2000, pp 983–988
44. Boscolo A, Quaia S, Tosto F (1998) No energy stored power conditioners for voltage sag compensation, mediterranean electrotechnical conference, MELECON 1998, pp 1194–1198

Chapter 2
Design of Taps Winding Structure and Configuration of Switches in Full-Electronic Tap-Changer

2.1 Introduction

Full-electronic tap-changer is faster, has better performance, more capabilities and lower maintenance costs compared with mechanical tap-changers and hybrid tap-changers; however, the cost of electronic tap-changers is higher than that of the conventional ones. This is one of the most important factors that prevents a wide application of this system [1]. Two cost deterministic factors in an electronic tap-changer are:

a. The number of solid-state power switches and their voltages and currents.
b. The number of transformer taps because every tap requires isolation equipment and its own specific insulators.

Therefore, if (a) and (b) decrease, the cost of tap-changers will be largely reduced. On the other hand, the higher the number of voltage steps over electronic tap-changer adjustment range, the more precise the regulation will be, and this is the most important characteristic of this system. Therefore, the major goals of the design of the power section of an electronic tap-changer are as follows [2]:

1. The number of solid-state power switches and their voltages and currents should be minimal.
2. The number of transformer taps should be minimal.
3. The number of voltage steps over output voltage regulation range should be maximal in order to enhance the regulation precision.

Of course, other lateral goals are also proposed and some of these goals are:

1. System losses including losses of switches, as well as transformer losses should be minimal.
2. System reliability should be high.

J. Faiz and B. Siahkolah, *Electronic Tap-changer for Distribution Transformers*,
Power Systems, DOI: 10.1007/978-3-642-19911-0_2,

3. Standing should be very larger than the rated current arising from the short circuit.
4. Standing should be very higher than the rated voltage due to the switching of the power system or lightning.

Although the above-mentioned lateral goals are very important, the major concentrations of this chapter are the major goals and the lateral goals will be proposed during arguments.

2.2 The Design of Transformer Taps Configuration

The first stage in the design of an electronic tap-changer is the design of transformer taps configuration. Certainly more reduction of the number of taps provides a better configuration while the number of voltage steps decreases.

In the first instance, if the diversity of the existing configurations for transformer taps is taken into account, the selection of optimal configuration does not seem straightforward. On the other hand, if it is claimed that a configuration is optimal, generally all possible configurations must be taken into account. Therefore, the optimal design of taps configuration in a transformer will be as such that first the number of the available configurations are reviewed; then, it will be tried to categorize comprehensively all possible configurations into a number of groups. Finally an optimal configuration is selected by studying the characteristics of each group.

Figure 2.1 exhibits some of the common configurations in mechanical, hybrid and also full-electronic tap-changers. Since this section concentrates on the taps winding configuration, only the transformer winding part has been drawn in this figure and switches arrangement has been presented as a black box. Meanwhile, the transformer phase has been shown and for a three-phase transformer similar configuration is repeated three times.

The taps in Fig. 2.1 may be placed in the primary or secondary, and this depends on some factors such as voltage and power of the transformer; however, for the proposed case, there is no difference between the two sides of the transformer. The terminals with hollow circle and inserted number 1 and 2 form two final terminals of the winding. N_is in Fig. 2.1 presents the main part of the winding. The main part of the winding is always in the circuit and in fact provides a constant bias for the winding turn number. The taps winding presented by n_i are used to regulate the turn number around bias turn number, and they can be in the circuit or independent of the switches position. Voltages of achievable cases are not equal, in other words, there is no repeated cases.

Considering the configurations of Fig. 2.1, they are categorized into three groups:

a. Those windings in which taps winding is placed on the main winding. In other words, taps windings are connected to the main winding internally (Fig. 2.1a and d).

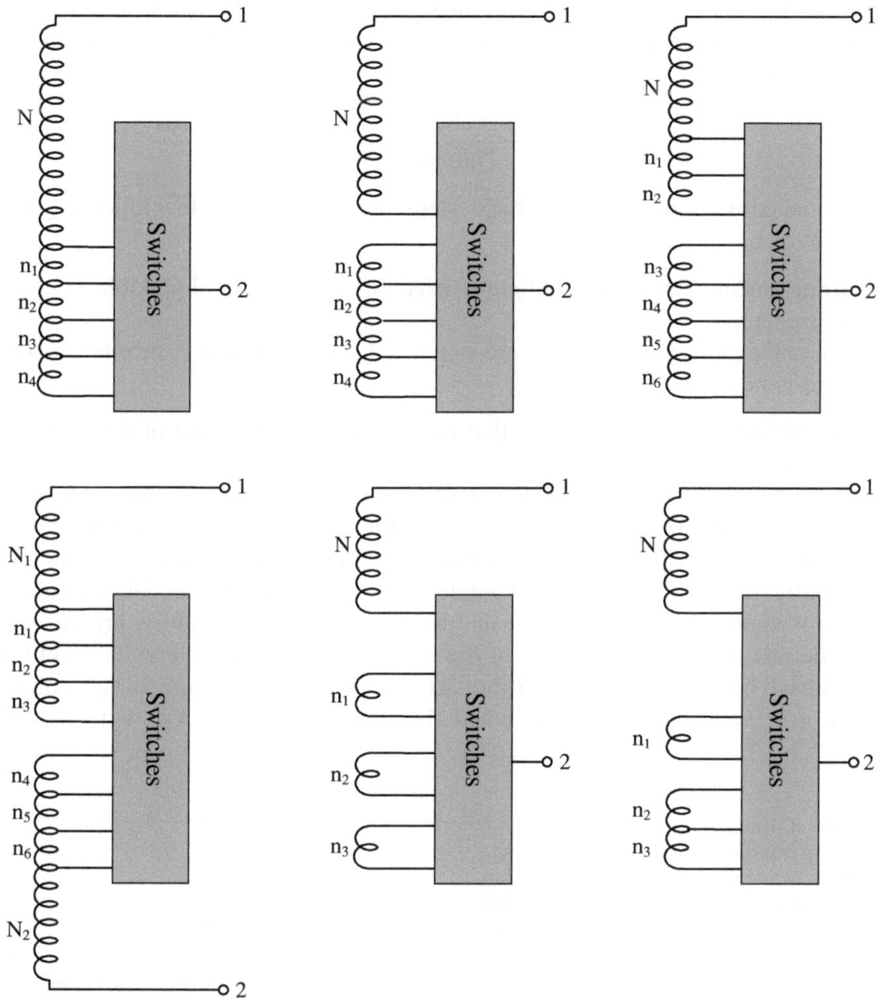

Fig. 2.1 Common configurations for taps winding in tapped-transformers and equipped by tap-changer

b. Those windings in which non of taps winding is place on the main winding, but place independently (Fig. 2.1b, e and f).
c. A combination of case a and b (Fig. 2.1c).

Meanwhile, only those voltages that do not cause short circuit in the windings are achievable.

Schematic of groups a and b for tapped-winding has been shown in Fig. 2.2. Number of taps in the configurations is equal to *K*.

Now, in order to compare these three groups, the following criterion C_1 is defined taking into account the main goals of the design for the proposed configuration:

$$C_1 = \frac{\text{Maximum achievable number of voltage steps}}{\text{Number of taps}} \tag{2.1}$$

The maximum achievable voltage steps are obtained by the following two assumptions:

a. Arrangement of switches due to taps selection as such that all possible cases are realizable.
b. The achievable voltage cases are not equal with each other, in other words, there is no repeated cases.

Meanwhile, only those voltage that do not cause short circuit of the windings are achievable.

Certainly, larger C_1 leads to a lower cost configuration which is more attractive. First C_1 is evaluated for configuration of Fig. 2.2a. Since transient modes of transformer are not proposed in this stage, the steady-state voltage is considered and configuration of Fig. 2.2a is modeled as Fig. 2.3. In this model, all voltage supply is sinusoidal and the phase and amplitude of each supply is proportional with the relevant number of turns of the winding. The achievable voltages in this model have been summarized in Table 2.1. As seen, V_{N1} is present as bias voltage in the achievable voltages. It is noted that voltages such as $V_{N1} + V_2$ are not

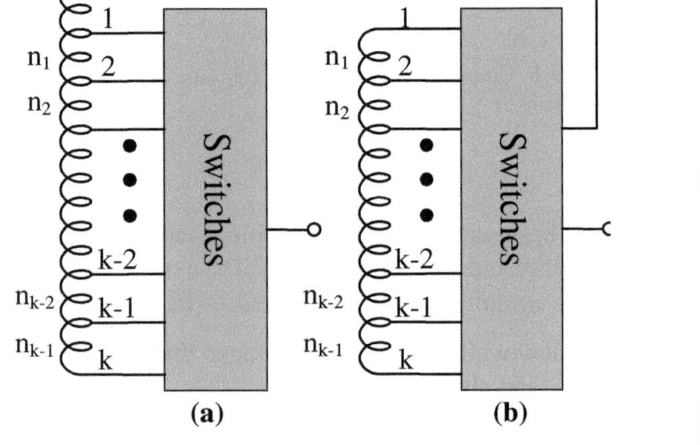

Fig. 2.2 Two different groups of windings for taps winding: **a** main winding connected to taps-winding, **b** main winding not connected to taps winding

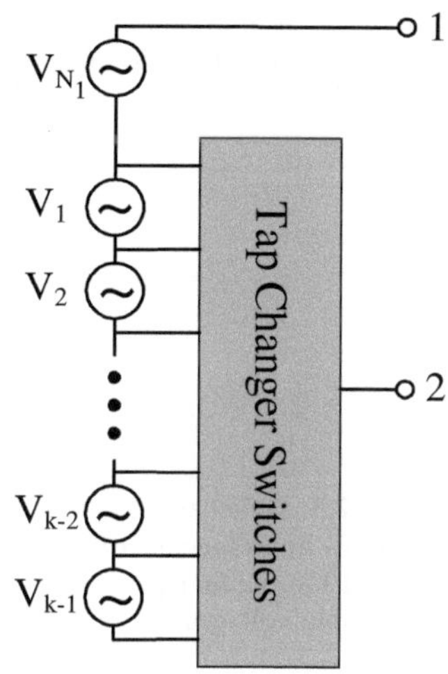

Fig. 2.3 Circuit model of configuration of Fig. 2.2a in steady-state

Table 2.1 Achievable voltages in model of Fig. 2.3

No.	Achievable voltage value	No. of values
1	V_{N_1}	1
2	$V_{N_1} + V_1$	1
3	$V_{N_1} + V_1 + V_2$	1
.	.	.
.	.	.
.	.	.
K–1	$V_{N1} + V_1 + V_2 + \cdots + V_{K-2}$	1
K	$V_{N1} + V_1 + V_2 + \cdots + V_{K-2} + V_{K-1}$	1
	Total	K

achievable in this model, because they cause short circuit of V_1. Therefore, criterion C_1 for configuration of Fig. 2.2a is calculated as follows:

$$C_1\big|_{\text{Configuration of Fig. 2.2a}} = \frac{K}{K} = 1 \tag{2.2}$$

Similarly, for configuration of Fig. 2.2b a corresponding circuit model with voltage supplies can be considered and then the achievable voltages are summarized in Table 2.2. There is V_{N1} as bias voltage over all values.

Referring to Table 2.3, sum of the achievable voltages for configuration of Fig. 2.2b is calculated as follows:

$$\begin{aligned}&\text{Sum of achievable voltages for configuration of Fig. 2.2b}\\&\quad = 1 + 2[(K-1) + (K-2) + \cdots + 2 + 1] = K^2 - K + 1\end{aligned} \tag{2.3}$$

Table 2.2 Achievable voltages for configuration of Fig. 2.2b

No.	Achievable voltage value	No. of values
1	V_{N_1}	1
2	$V_{N_1} + V_1$	2(K−1)
3	$V_{N1} + V_1 + V_2$	2(K−2)
.	.	.
.	.	.
.	.	.
K−1	$V_{N1} + V_1 + V_2 + \cdots + V_{K-2}$	2
K	$V_{N1} + V_1 + V_2 + \cdots + V_{K-2} + V_{k-1}$	1
	Total	$K^2 - K + 1$

Table 2.3 Switches positions in arrangement of Fig. 2.10 to obtain each achievable voltages

No.	Switches positions					Value of realized voltage
	S_1	S_2	$\cdots$	S_{K-1}	S_K	
1	1	0	$\cdots$	0	0	V_{N_1}
2	0	1	$\cdots$	0	0	$V_{N_1} + V_{n_1}$
•						
•						
•						
K	0	0	$\cdots$		1	$V_{N_1} + V_{n_1} + V_{n_2} + \cdots + V_{n_{K-1}}$

Therefore, criterion C_1 for configuration of Fig. 2.2b is obtained as follows:

$$C_1|_{\text{Configuration of Fig. 2.2b}} = \frac{K^2 - K + 1}{K}; \quad K \geq 2 \tag{2.4}$$

Figure 2.5a is the result of Fig. 2.5b repetition and Fig. 2.5b the result of Fig. 2.2b repetition. If sum of taps in configuration of Fig. 2.5 is represented by T, it can be written as

$$T = (K_1 + 1) + (K_2 + 1) + \cdots + (K_1 + 1) \tag{2.5}$$

Now this question can be proposed: if total taps are a given constant value of T, for which values of K_1, K_2, ..., K_i, is criterion C_1 maximized in each of the configurations of Fig. 2.5a, b? Before to answer this question, it must be noted that in the configurations having several tapped-windings, the maximum number of the achievable voltages are obtained from the product of the achievable voltages due to one by one windings; in other words, if the number of achievable voltages of the tapped-winding is represented by N_V, the total number of the achievable voltages $N_{v\text{Total}}$ is obtained as follows:

$$N_{V\text{Total}} = N_{V_1} N_{V_2} \ldots N_{Vi} = j = 1 \underset{\Sigma}{\overset{i}{\Pi}} N_{V_j} \tag{2.6}$$

where N_{V_j} is the number of achievable voltages due to the jth tapped-winding. Now the above-mentioned question can be answered by referring to this introduction. First configuration of Fig. 2.5a is investigated. Criterion C_1 for this configuration is obtained as follows:

$$C_1|_{\text{Configuration of Fig. 2.5a}} = \frac{(K_1+1)(K_2+1)\cdots(K_i+1)}{(K_1+1)+(K_2+1)+\cdots+(K_i+1)} \tag{2.7}$$

Since it has been assumed that the given total taps is a constant value of T, considering that when the sum of several positive numbers is constant, their product is maximized when all the numbers are equal. The maximum value of C_1 will be as follows:

$$C_1|_{\text{Configuration of Fig. 2.5a}} = \frac{\left(\frac{T}{i}\right)^i}{T} \tag{2.8}$$

where i is the number of tapped-windings. Figure 2.4 compares criterion C_1 for configurations of Fig. 2.2a, b for equal taps number. Now it must be investigated which value of i maximizes Eq. 2.8. At this end, Eq. 2.8 is rewritten as follows:

$$C_1|_{\text{Configuration of Fig. 2.5a}} = \frac{(n)^{\frac{T}{n}}}{T} = \frac{\left(n^{\frac{1}{n}}\right)^T}{T} \tag{2.9}$$

$n = T/i$ and is equal to the number of taps in each tapped-windings. Since T, n and $n^{1/n}$ are equal or larger than unity, Eq. 2.9 is maximized when $n^{1/n}$ is maximized. This function has been plotted in Fig. 2.6 in which is maximized at $n = 3$. It is concluded that configuration of Fig. 2.5a for a fixed number of taps, is maximized when it has the maximum number of the achievable voltages where all its tapped-windings have three taps. In this case the peak value of C_1 is as follows:

$$C_{1\text{Max}} - \frac{\left(3^{\frac{1}{3}}\right)^T}{T} = \frac{1.44^T}{T} \tag{2.10}$$

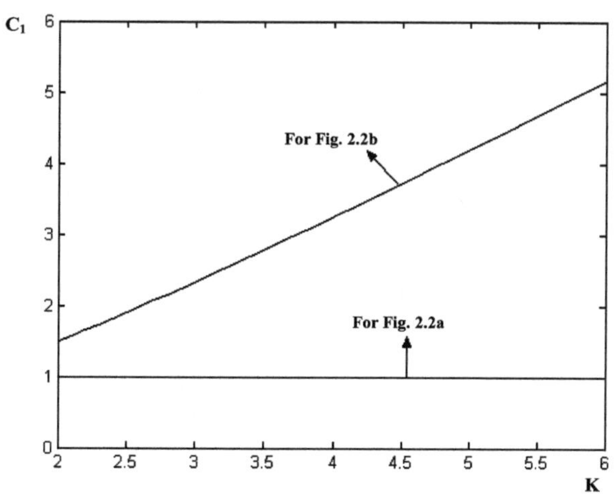

Fig. 2.4 C_1 versus number of taps (K) for configurations of Fig. 2.2a and b

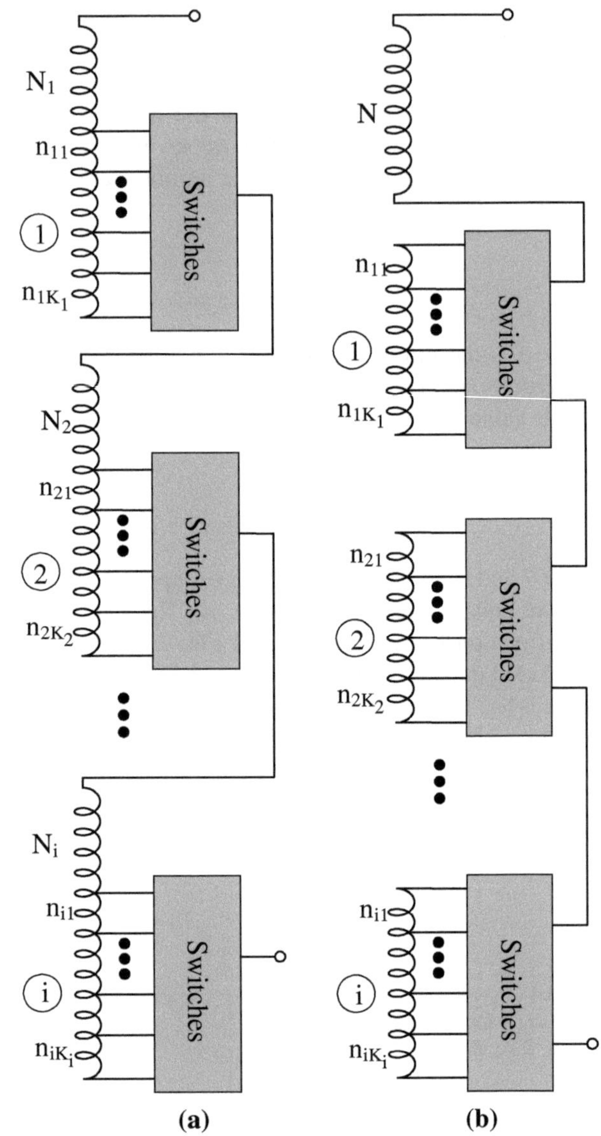

Fig. 2.5 Configuration of taps winding when there are several taps: repetition of configuration of a. Fig. 2.2a and b

Now configuration of Fig. 2.5b is investigated. Criterion C_1 for this configuration is obtained as follows:

$$C_1\Big|_{\text{Configuration of Fig. 2.5b}} = \frac{(K_1^2 - K_1 + 1)(K_2^2 - K_2 + 1)\cdots(K_i^2 - K_i + 1)}{(K_1 + 1) + (K_2 + 1) + (K_3 + 1) + \cdots + (K_i + 1)} \tag{2.11}$$

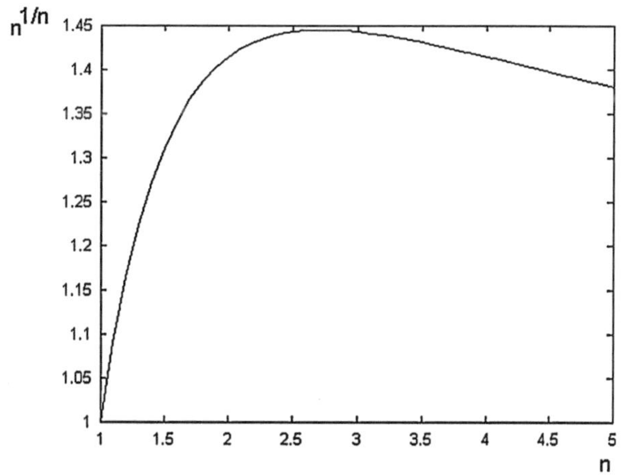

Fig. 2.6 Variations of $n^{1/n}$ versus n

In this case, it can be shown again that Eq. 2.11 is maximized if all tapped-windings have an equal number of taps. For number of taps equal to n:

$$C_{1\,\mathrm{Max}}\big|_{\text{Configuration of Fig. 2.5b}} = \frac{(n^2 - n + 1)^{\frac{T}{n}}}{T} \tag{2.12}$$

Equation 2.12 will be maximized when $(n^2 - n + 1)^{1/n}$ is maximized. Figure 2.7 shows the variations of the mentioned function against n. As seen this function is maximized at $n = 3$. Consequently, for a fixed number of taps, configuration of Fig. 2.5b has the peak values of the achievable voltages if all tapped-windings have three taps. In this case, the maximum value of the criterion will be as follows:

$$C_{1\,\mathrm{Max}}\big|_{\text{Configuration of Fig. 2.5b}} = \frac{\left((3^2 - 3 + 1)^{\frac{1}{3}}\right)^T}{T} = \frac{1/91^T}{T} \tag{2.13}$$

Comparison of Eqs. 2.10 and 2.13 indicates that the configuration of Fig. 2.5b has the larger maximum criterion C_1 than that of Fig. 2.5a. Combination of the configuration of Fig. 2.5a and b is also possible (like Fig. 2.1c). However, for criterion C_1 the configuration of Fig. 2.5b will be certainly optimal because if each of the tapped-windings of Fig. 2.2a and tapped-windings of Fig. 2.2b are replaced with the number of taps, number of the achievable voltages will be higher (according to Fig. 2.4). Therefore, the number of the achievable voltages will also be larger.

Therefore, the optimal configuration for taps windings from C_1 criterion point of view will be as shown in Fig. 2.8. Total number of taps in this configuration is $T = 3i$ and the C_1 criterion is: $V^i/3i$.

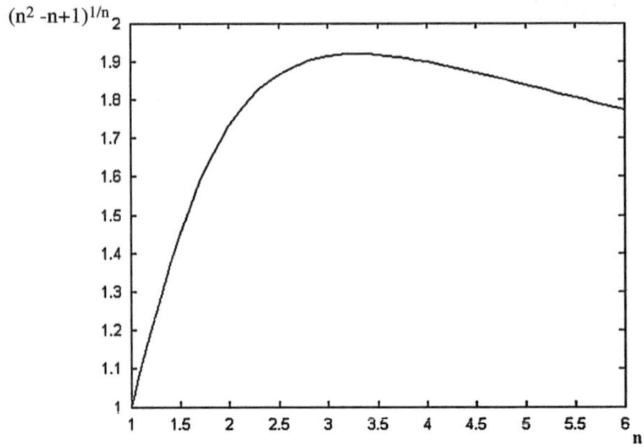

Fig. 2.7 Variation of $(n^2 - n + 1)^{1/n}$ versus n

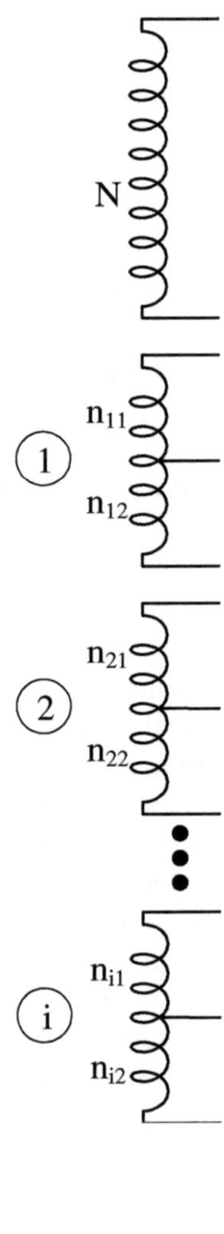

Fig. 2.8 Optimal configuration for windings using criterion $C1$

2.3 Voltage Selection of Each Part of Windings in Tapped-Winding

It was mentioned in Sect. 2.2 that the number of the achievable voltages is evaluated assuming unequal voltages. Therefore, the voltage of each section of the windings in a tapped-winding must be chosen as such that the achievable cases are not repetitive. Meanwhile, in most applications, it is necessary that the windings voltage be regulated as such that the achievable voltages are linear with equal step voltages. In this section, the voltages of each section of windings are determined for the configuration of Fig. 2.5 and optimal case, and the equal taps number in any tapped-winding (three taps in each tapped-winding). It is again emphasized that the choice of voltages is based on the two above-mentioned conditions.

The first configuration of Fig. 2.5a is considered. Figure 2.9 exhibits the configuration of Fig. 2.5a for the number equal to three in each tapped-winding, where

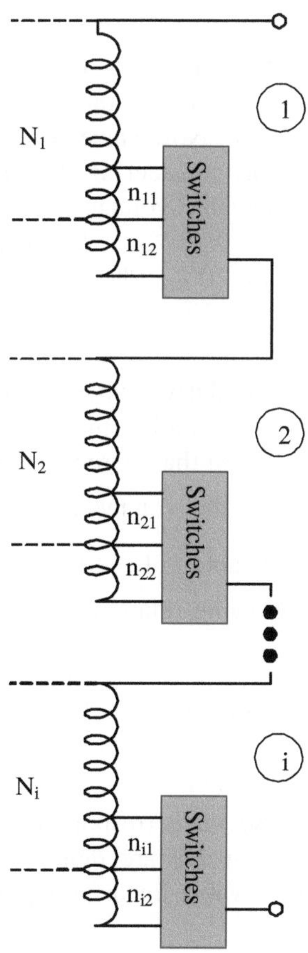

Fig. 2.9 Configuration of Fig. 2.2a for $n = 3$

in each tapped-winding, the number of bias turns up to middle tap is shown by N_j and windings of taps by n_{j_1}, n_{j_2}.

The achievable voltages in configuration of 2.9 are as follows:

$$\text{Achievable voltages in configuration of Fig. 2.9} = V_{N_1} + V_{N_2} + \cdots + V_{N_i} + \begin{matrix} - & V_{n_{11}} & - & V_{n_{21}} & \cdots & - & V_{n_{i1}} \\ & 0 & + & 0 & \cdots & + & 0 \\ + & V_{n_{12}} & + & V_{n_{22}} & \cdots & + & V_{n_{i2}} \end{matrix} \tag{2.14}$$

where V_X corresponds with winding X. It is clear that if the achievable voltages are placed around bias voltage symmetrically, the following relationship must be satisfied:

$$V_{n_{11}} = V_{n_{12}},\ V_{n_{21}} = V_{n_{22}}, \ldots, V_{n_{i1}} = V_{n_{i2}} \tag{2.15}$$

If step voltage between the achievable voltages is called V_{step}, then, it is clear that at least one of the taps windings (for example the first winding) must have a voltage equal to V_{step}; in other words, the following must be satisfied:

$$V_{n_{11}} = V_{n_{12}} = V_{\text{step}} \tag{2.16}$$

So, voltages $-V_{\text{step}}$, 0, V_{step} will be realized around the bias voltage (to simplify, hereafter the bias voltage as a constant and definite voltage is not mentioned). Now values of $V_{n_{12}}$ and $V_{n_{22}}$ are determined. If repetitive voltages do not occur, considering the second tapped-winding, in total there will be $3 \times 3 = 9$ achievable voltages. If these achievable voltages differ by equal steps V_{step}, the cases 0, $\pm V_{\text{step}}$, $\pm 2V_{\text{step}}$, $\pm 3V_{\text{step}}$, and $\pm 4V_{\text{step}}$ will be realized. Since $V_{n_{21}} = V_{n_{22}} = 3V_{\text{step}}$, the following must be satisfied:

$$V_{n_{21}} = V_{n_{22}} = 3V_{\text{step}} \tag{2.17}$$

Similarly, voltages for windings n_{j_1} and n_{j_2} can be determined. Thus, for jth tapped-winding the total number of the achievable voltages will be 3^j. If it is assumed that there is no repetitive voltages and the achievable voltages have equal steps, the voltage range of $\pm\left[\frac{3^j-1}{2}\right]V_{\text{step}}$ will be realizable, while there will be realizable voltage range of $\pm\left[\frac{3^{j-1}-1}{2}\right]V_{\text{step}}$ up to winding (j – 1)th. Therefore, the following relationship must be satisfied:

$$V_{n_{j1}} = V_{n_{j2}} = \left[\frac{3^j - 1}{2}\right]V_{\text{step}} - \left[\frac{3^{j-1} - 1}{2}\right]V_{\text{step}} = 3^{j-1}V_{\text{step}}; \quad j = 1, 2, \ldots, i \tag{2.18}$$

So, the corresponding voltages of each section of taps windings in configuration of Fig. 2.9 are determined to prevent repetitive voltage and also to provide achievable linear voltages with equal steps of V_{step}.

Now consider configuration of Fig. 2.5b. This configuration is presented in Fig. 2.8 for three taps in each of the tapped-windings. The achievable voltages in configuration of Fig. 2.8 are as follows:

$$
\text{Achievable voltages in configuration of Fig. 2.8}
$$

$$
= V_N + \begin{array}{cccccc}
- & (V_{n_{11}} + V_{n_{12}}) & - & \cdots & - & (V_{n_{i_1}} + V_{n_{i_2}}) \\
- & V_{n_{12}} & - & \cdots & - & V_{n_{i_2}} \\
- & V_{n_{11}} & - & \cdots & - & V_{n_{i_1}} \\
 & 0 & + & \cdots & + & 0 \\
+ & V_{n_{11}} & + & \cdots & + & V_{n_{i_1}} \\
+ & V_{n_{12}} & + & \cdots & + & V_{n_{i_2}} \\
+ & (V_{n_{11}} + V_{n_{12}}) & + & \cdots & + & (V_{n_{i_1}} + V_{n_{i_2}})
\end{array} \tag{2.19}
$$

It is clear from Eq. 2.17 that the voltage selected for any taps winding, the achievable voltages around bias voltage (V_N) will be symmetrical and no caution is required. However, in order to have equal achievable voltages of V_{step}, the following relationship must be at least satisfied:

$$
V_{n_{11}} = V_{\text{step}}, \; V_{n_{12}} = 2V_{\text{step}} \tag{2.20}
$$

Also, the following must be satisfied in all tapped-windings:

$$
2V_{n_{j1}} = V_{n_{j2}}; \quad j = 1, 2\ldots, i \tag{2.21}
$$

To determine $V_{n_{j1}}$ and $V_{n_{j2}}$, for ith tapped-winding the total number of achievable voltages is V_j. Suppose there is no repetitive voltage, and the achievable voltage has equal steps. In this case there will be the realizable voltages up to the range of $\pm\left(\frac{7^i-1}{2}\right)V_{\text{step}}$ for jth tapped-winding while these voltages up to winding $(j-1)$th will be $\pm\left[\frac{V^{j-1}-1}{2}\right]V_{\text{step}}$. Therefore:

$$
V_{n_{j1}} + V_{n_{j2}} = \left(\frac{7^j - 1}{2}\right)V_{\text{step}} - \left(\frac{7^{j-1} - 1}{2}\right)V_{\text{step}} = 3 \times 7^{j-1}V_{\text{step}} \tag{2.22}
$$

Referring to Eq. 2.21;

$$
V_{n_{j1}} + 2V_{n_{j1}} = 3 \times 7^{j-1}V_{\text{step}} \tag{2.23}
$$

Hence, $V_{n_{j1}}$ and $V_{n_{j2}}$, obtained as follows:

$$
V_{n_{j1}} = 7^{j-1}V_{\text{step}}; \quad V_{n_{j2}} = 2 \times 7^{j-1}V_{\text{step}}; \quad j = 1, 2\ldots i \tag{2.24}
$$

Therefore, the corresponding voltages (V_{step}) of each section of the taps windings in configuration of Fig. 2.8 are determined as such that there is no repetitive voltages and the linear achievable voltages with equal steps are provided.

2.4 Switches Arrangement

One of the assumptions in Sect. 2.2 for evaluation of the maximum number of achievable voltage was that the switches arrangement must be such that all achievable voltages are realized. This is proposed in this section. First arrangement of Fig. 2.2a is considered. Fig. 2.10 presents the switches arrangement in which tapped-winding realizes all possible cases.

As seen, in this switches arrangement, the number of switches is equal to the number of taps and in order to obtain the achievable voltages, in any case one of the switches must be on and others off. Table 2.3 presents the position of switches for obtaining the realizable voltages.

Now configuration of Fig. 2.2b is considered. Figure 2.11 suggests four switches arrangements for this configuration. Every suggested arrangements, in Fig. 2.11 can realize a specific number of the achievable voltages. To compare these arrangements the following criterion is defined:

$$C_2|_{\text{specific configuration of Fig. 2.11a}} = \frac{\text{No. of realizable voltages}}{\text{No. of switches}} \tag{2.25}$$

A larger C_2 leads more suitable arrangements. Now criterion C_2 is calculated for each arrangement of Fig. 2.11. For arrangement of Fig. 2.11a, criterion C_2 can be easily evaluated as follows:

$$C_2|_{\text{Configuration of Fig. 2.10a}} = \frac{K}{K} = 1 \tag{2.26}$$

where K is the number of taps. It is clear that arrangement of Fig. 2.11b can realize all achievable voltages, so criterion C_2 for this arrangement will be as follows:

$$C_2|_{\text{Configuration of Fig. 2.11b}} = \frac{K^2 - K + 1}{2K} \tag{2.27}$$

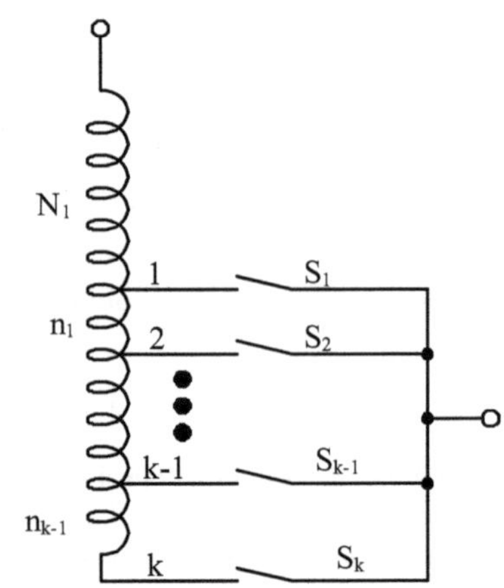

Fig. 2.10 Switches arrangement for configuration of Fig. 2.2a

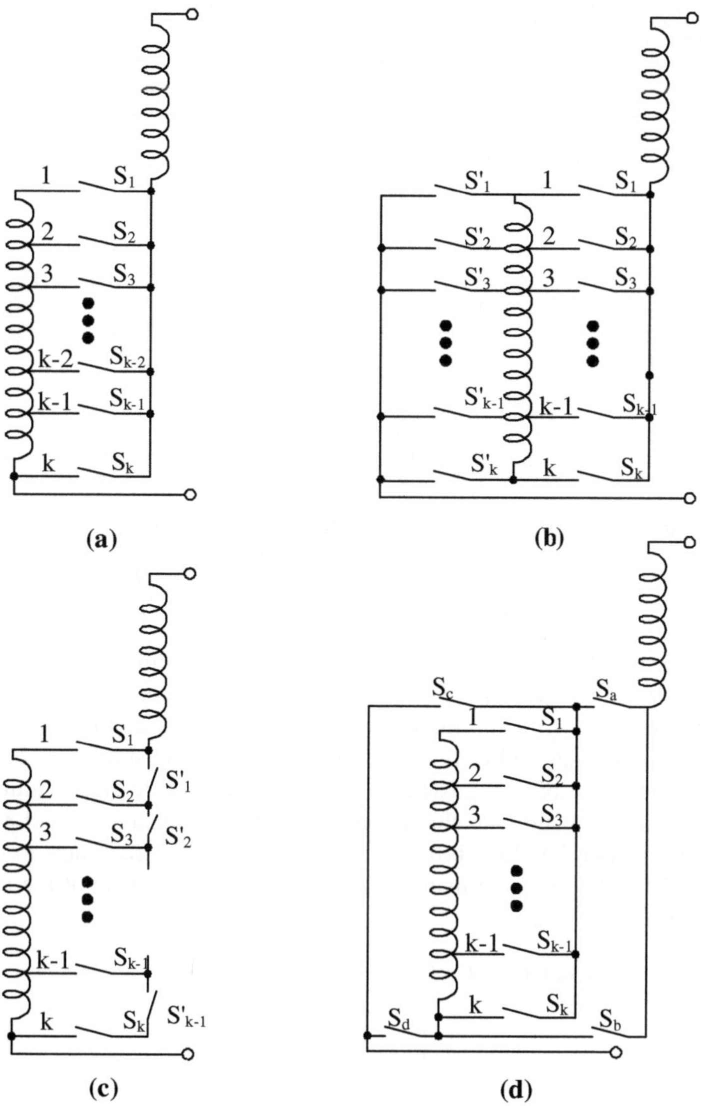

Fig. 2.11 Four types of suggested switches arrangements for configuration of Fig. 2.2b

Table 2.4 summarizes the switches arrangements of Fig. 2.11b to realize the achievable voltages.

Arrangement of Fig. 2.11c realizes all achievable voltages in Table 2.2 with + sign, therefore criterion C_2 of this arrangement will be as follows:

$$C_2|_{\text{Configuration of Fig. 2.11c}} = \frac{\frac{[K^2-K+1]-1}{2}+1}{2(K-1)+1} = \frac{\frac{K(K-1)}{2}+1}{2K-1} \tag{2.28}$$

Table 2.4 Switches arrangements of Fig. 2.11b to realize achievable voltages

Switches positions										Value of realized voltage
S_1	S_2	$\cdots$	S_{K-1}	S_K	$S_1^{'}$	$S_2^{'}$	$\cdots$	$S_{K-1}^{'}$	$S_K^{'}$	
1	0	$\cdots$	0	0	0	1	0	0	0	V_{n_1}
0	1	$\cdots$	0	0	0	0	1	0	0	V_{n_2}
.	.	.	.	.	.	.		.	.	.
	.	.	.	.	.	.		.	.	.
.	.	.	.	.	.	.		.	.	.
0			1	0	0				1	$V_{n_{k-1}}$
0	1	$\cdots$		0	1	0		0	0	$-V_{n_1}$
0	0	1		0	0	1		0	0	$-V_{n_2}$
.	.	.	.	.	.	.		.	.	.
.	.	.	.	.	.	.		.	.	.
.	.	.	.	.	.	.		.	.	.
0	0	0		1	0	0		1	0	$-V_{n_{k-1}}$

Finally arrangement of Fig. 2.11d has the following criterion:

$$C_2|_{\text{Configuration of Fig. 2.11d}} = \frac{2(K-1)+1}{K+4} = \frac{2K-1}{K+4} \tag{2.29}$$

For $K \geq 3$ Fig. 2.11b is more suitable than other arrangements from criterion C_2 point of view. Arrangement of Fig. 2.11b is an arrangement that realizes all achievable voltages with minimum switches. Figure 2.12 compares C_2 criterion for configuration suggested in Fig. 2.11. To prove this, consider Fig. 2.13. In this figure, switches have been represented as a black box where taps winding are the input and terminals I and II as the output of the box. Realization of all achievable voltages means that the black box must be able to connect each taps to terminal I and terminal II. Meanwhile, this must be done as such that none of the taps short-circuited with each other.

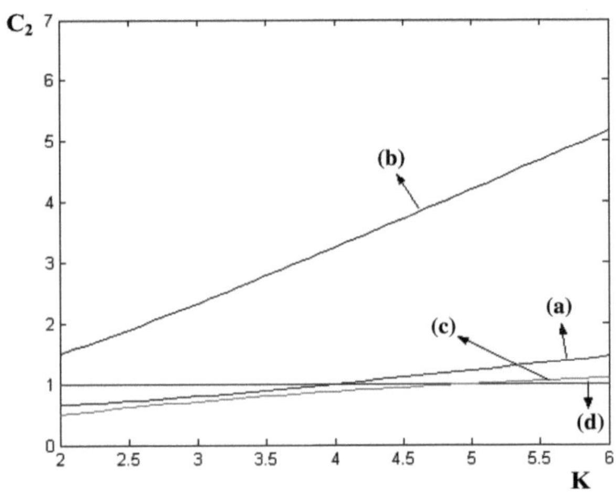

Fig. 2.12 Criterion curve for arrangements of Fig. 2.11

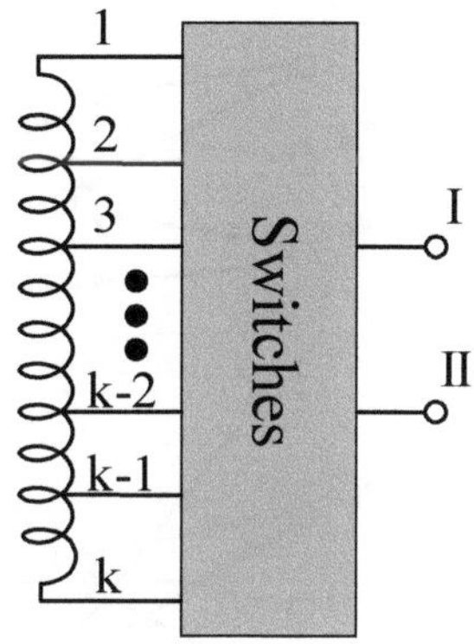

Fig. 2.13 Representing switches arrangements as a black box-winding taps as input and terminals I and II as output of black box

It is necessary that every winding taps be connected to terminal I and also each tap is connected to terminal II by independent switches. Such arrangement has been shown in Fig. 2.14. A close look at this arrangement indicates that this arrangement is identical with that of Fig. 2.11b which is better than other arrangement from the criterion C_2 point of view. Thus, at least twice of the taps numbers are required to realize all achievable voltages.

So far, it has been shown that arrangement of Fig. 2.11b has a larger criterion C_2 among the suggested arrangements in Fig. 2.11, However, it has not yet been proved whether from criterion C_2 point of view, arrangement of Fig. 2.11b is an optimal arrangement among all possible arrangements. It will be shown that the above-mentioned arrangement is optimal from C_2 criterion point of view among all possible arrangements. At this end, block box of Fig. 2.13 is drawn as Fig. 2.15. In this figure each tap has been shown by a node (circle) with numbers n_1–n_K and the output terminals with nodes I and II.

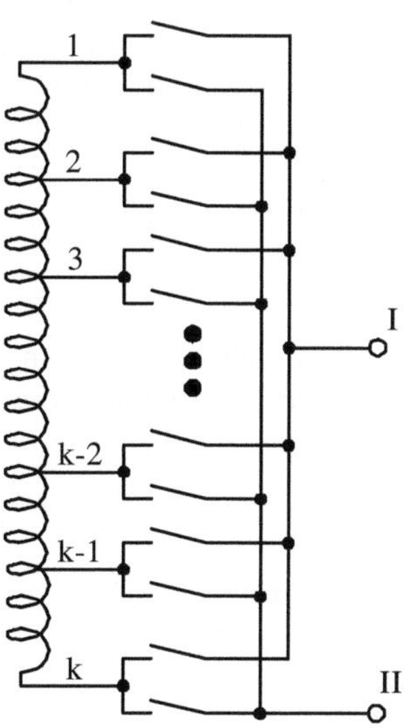

Fig. 2.14 An arrangement realizing all achievable voltages

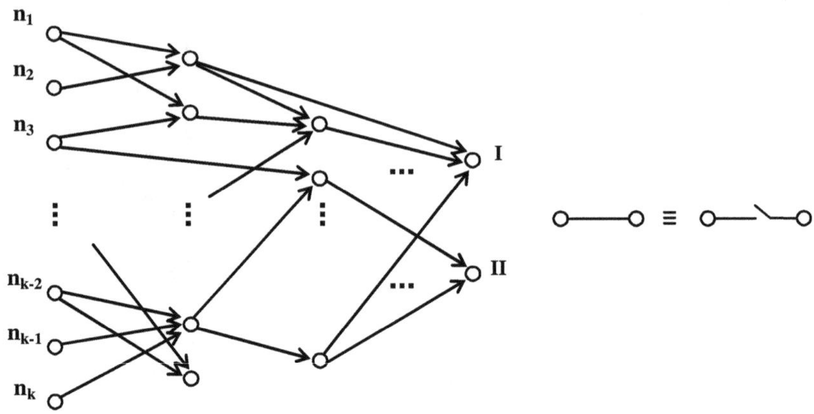

Fig. 2.15 General arrangement for switches

Meanwhile, the number of middle nodes is also considered. Switches have been shown by the lines is drawn between the nodes. At least two switches enter and one switch exits from every middle node. There is not any restriction in connecting switches. Fig. 2.15 presents all possible arrangements for tap-changing configuration of Fig. 2.2b. Now criterion C_2 is evaluated for arrangement of Fig. 2.15. It must be emphasized that the realizable voltages in this general arrangement are those in which there is no any common middle between connection of node I to taps and node II to their taps. Otherwise, the taps will be short-circuited. Suppose that the general arrangement of Fig. 2.15 is able to connect terminal I to J taps and for the connection of terminal I to tap ith (i = 1, 2, ..., J), terminal II will be able to connect to other T_i taps from independent middle nodes. So, when terminal I is able to connect to J different taps, then at least there need for J switches for such a task. On the other hand, when it is possible to connect terminal II to other T_i taps for connecting terminal I to ith tap, T_1 independent switches must exist. In total, if terminal II can be connected to T_t taps for different cases. Then, the following can be certainly written:

$$T_t \geq T_i; \quad i = 1, 2, \ldots, J \tag{2.30}$$

It is concluded that there will be at least T_1 independent switches to connect terminal II to this number of taps. Thus, criterion C_2 is as follows:

$$C_2 = \frac{1 + \sum_{i=1}^{J} T_i}{1 + J + T_t}; \quad 1 < J \leq K, \; 1 < T_t \leq K - 1 \tag{2.31}$$

Regarding Eq. 2.30:

$$C_2 \leq \frac{JT_t + 1}{J + T + 1_t} \tag{2.32}$$

It is clear that function $\frac{JT_t}{J+T_t}$ is maximized when J and T_t are equal and identical with their maximum possible value K, therefore:

$$C_2 \leq \frac{K^2 - K + 1}{2K} \tag{2.33}$$

2.5 Optimal Topology Exclusive of Influence of Voltage Amplitude and Switches Currents

A tapped-windings configuration with the minimum number of taps and maximum achievable voltages were introduced in Sect. 2.2. As shown in Sect. 2.4, the optimal arrangement of switches with minimum number of switches realize all achievable voltages. Figure 2.16 presents optimal structure of the taps with optimal arrangement of switches.

There are the following relationships in Fig. 2.16:

$$N_T = 3i \tag{2.34}$$

$$N_S = 6i \tag{2.35}$$

$$N_V = 7^i \tag{2.36}$$

$$2V_{n_{j1}} = V_{n_{j2}} = 2V_{\text{step}}7^{j-1}; \quad j = 1, 2, \ldots, \text{I} \tag{2.37}$$

where N_r is the taps number, N_s is the number of switches, N_v is the number of achievable voltages and $V_{n_s}$is the voltage corresponding with winding n_s.

The number of taps and number of current and voltage of the solid-state switches are those factors that determine the cost of the electronic tap-changer. In this section the voltage and current impacts are ignored and only their numbers are taken into account. So the cost function can be given as follows:

$$C_3 = \alpha_{N_T}N_T + \alpha_{N_S}N_S \tag{2.38}$$

where α_{N_T} is the weight related to the cost of each tap and α_{N_s} is the weight related to each switch. Is the topology shown in Fig. 2.16 is optimal from the optimal C_3 criterion point of view?. In other words, if a tap-changer with N_v realizable voltages are desired, which topology has the minimum cost (cost function of 2.28, and assume α_{N_T} and α_{N_s} are known)?

It is clear that the optimal topology is a tap-changer shown in Fig. 2.16 or shown in Fig. 2.17. To find the optimal topology, the cost function C_3 is calculated for two suggested topologies.

First the cost function C_3 is calculated for topology of Fig. 2.16. Referring to Eqs. 2.34–2.36, N_T and V_S versus N_v is substituted in Eq. 2.38 and the following equation is obtained:

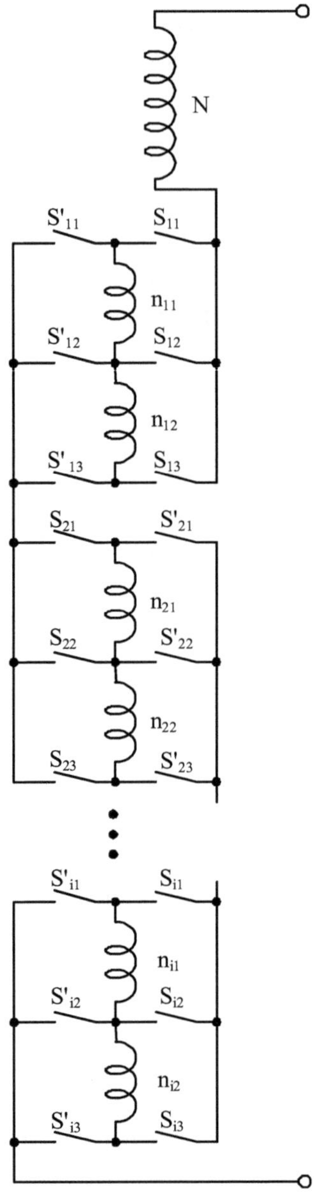

Fig. 2.16 Optimal structure of the taps with optimal arrangement of switches

$$\begin{aligned} C_3\big|_{\text{Topology of Fig. 2.16}} &= \alpha_{N_T} \times 3\log_7 N_V + \alpha_{N_S} \times 6\log_7 N_V \\ &= 3\log_7 N_V(\alpha_{N_T} + 2\alpha_{N_S}) \end{aligned} \tag{2.39}$$

Similarly, the cost function C_3 is obtained for topology of Fig. 2.17:

$$C_3\big|_{\text{Topology of Fig. 2.17}} = 3(\alpha_{N_T} + \alpha_{N_S})\log_3 N_V \tag{2.40}$$

Fig. 2.17 Candidate topology for optimal topology based on cost function C_3

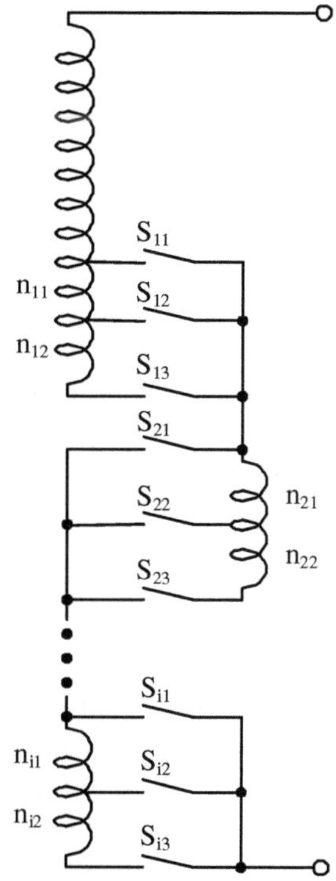

Now, to compare the cost related to the proposed topologies, the ratio of this function is studied. The following equation is the ratio of the relevant cost function of Fig. (2.16), and cost function related to (2.17):

$$\frac{C_3|_{\text{Topology of Fig. 2.17}}}{C_3|_{\text{Topology of Fig. 2.16}}} = \frac{3\log_7 N_V(\alpha_{N_T} + 2\alpha_{N_S})}{3\log_3 N_V(\alpha_{N_T} + \alpha_{N_S})} = \log_7 3 \times \frac{1 + 2(\alpha_{N_S}/\alpha_{N_T})}{1 + (\alpha_{N_S}/\alpha_{N_T})}$$
$$= 0/5646 \times \frac{1 + 2(\alpha_{N_S}/\alpha_{N_T})}{1 + (\alpha_{N_S}/\alpha_{N_T})} \tag{2.41}$$

If the above ratio is larger than unity, topology of Fig. 2.17 will be optimal. If it is smaller than unity, topology 2.16 will be optimal. This ratio depends on the ratio of $\alpha_{N_s}/\alpha_{N_T}$, and this in turn depends upon the tap-changer power and its voltage and the side (primary or secondary) of the placing taps. Equations 2.42 and 2.43 define range of the ratio $\alpha_{N_s}/\alpha_{N_T}$, in which the topologies of of Figs. 2.16 and 2.17 are optimal, respectively. Meanwhile, Fig. 2.18 presents the curve of the ratio of the cost functions.

$$0/5646 \times \frac{1 + 2(\alpha_{N_S}/\alpha_{N_T})}{1 + (\alpha_{N_S}/\alpha_{N_T})} < 1 \Rightarrow \frac{\alpha_{N_S}}{\alpha_{N_T}} < \frac{0/4354}{0/1292} = 3/37 \tag{2.42}$$

$$0/5646 \times \frac{1 + 2(\alpha_{N_S}/\alpha_{N_T})}{1 + (\alpha_{N_S}/\alpha_{N_T})} > 1 \Rightarrow \frac{\alpha_{N_S}}{\alpha_{N_T}} > 3/37 \tag{2.43}$$

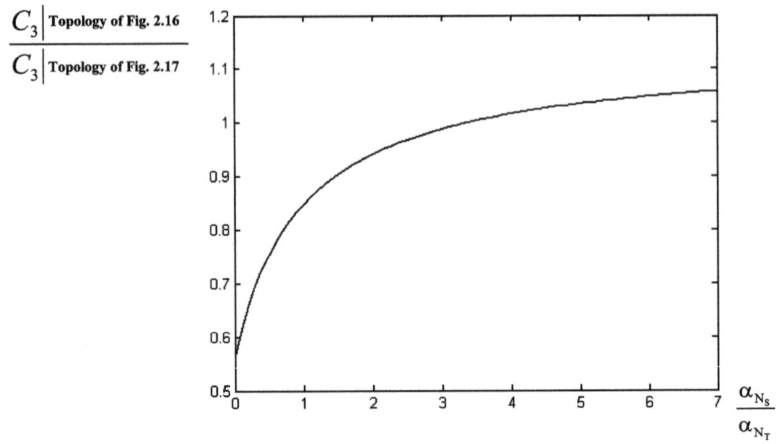

Fig. 2.18 Curve of ratio of cost functions of topologies of Figs. 2.16 and 2.17 versus α_{Ns}/α_{N_T} ratio

From Eqs. 2.42 and 2.43, it is concluded that if the cost of adding a switch is <3.37 times of the cost of adding a tap, then this is not necessary, the topology of Fig. 2.16 will be optimal based on the cost function C_3. If the cost of adding a switch is higher than 3.37 times of the cost of adding a tap, then the topology of Fig. 2.17 will be optimal again based on the cost function C_3.

2.6 Optimal Topology Considering Impact of Voltages and Currents of Switches

In this section the cost function is defined as such that it includes the impact of the voltages and currents of the switches. Of course, in all proposed topologies, the current of each tap-changer switches in connection will be identical with the rated current of the transformer; therefore, it is not necessary to consider the currents of the switches for comparison of different topologies. The following cost function includes the impacts of the voltages of the switches:

$$C_4 = \alpha_{N_T} N_T + \alpha_{\text{V}} \sum_{j=1}^{N_S} V_j \tag{2.44}$$

Equation 2.44 assumes that the cost of each switch is linearly proportional with its voltage, and α_v is the related factor.

Before to calculate cost function C_3 for topology of Figs. 2.16 and 2.17, the peak voltage of the terminals of each switch is expressed in Fig. 2.19. In this figure, V_1 and V_2 are the rms voltages of the windings and it is assumed that at least one switch connects in each case.

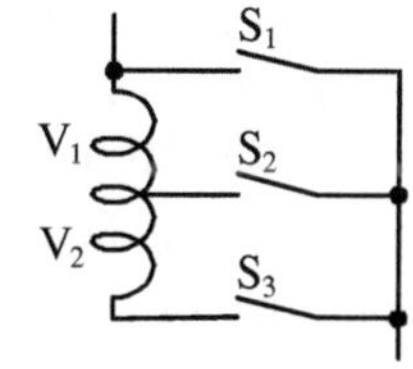

Fig. 2.19 A typical tapped-winding with its switches

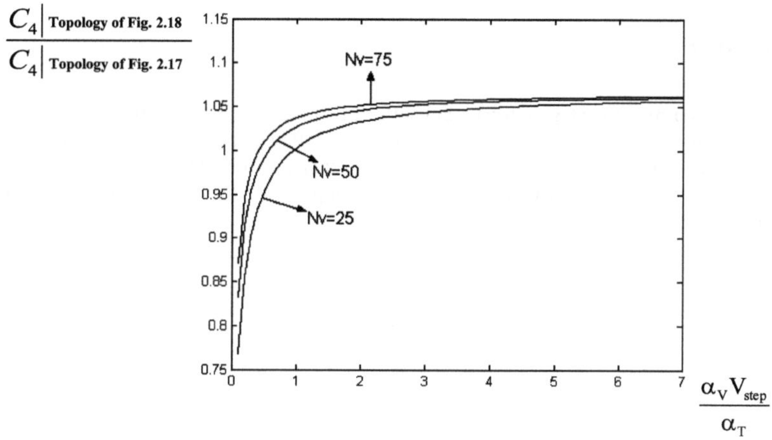

Fig. 2.20 Ratio of the two cost functions versus α_{Ns}/α_{N_T} for different N_V

$$\text{Peak voltage at terminals of switch } S_1 = \sqrt{2}(V_1 + V_2)$$
$$\text{Peak voltage at terminals of switch } S_2 = \sqrt{2}\,\text{Max}\{V_1 + V_2\}$$
$$\text{Peak voltage at terminals of switch } S_3 = \sqrt{2}(V_1 + V_2)$$

By this introduction, cost function C_4 is calculated for each of the topologies (Figs. 2.16 and 2.17) as follows:

$$\begin{aligned} C_4\big|_{\text{Topology of Fig. 2.16}} &= 3\alpha_{N_T}\log_7 N_V + \alpha_V \sum_{j=1}^{\log_7 N_V} \left(2 \times 2V_{\text{step}}V^{j-1} + 4 \times 3V_{\text{step}}V^{j-1}\right) \\ &= 3\alpha_{N_T}\log_7 N_V + \alpha_V \times 16V_{\text{step}}\left(\frac{N_V - 1}{6}\right) \end{aligned} \tag{2.45}$$

$$\begin{aligned} C_4\big|_{\text{Topology of Fig. 2.17}} &= 3\alpha_{N_T}\log_3 N_V + \alpha_V \sum_{j=1}^{\log_3 N_V} \left(1 \times 1V_{\text{step}}3^{j-1} + 2 \times 2V_{\text{step}}3^{j-1}\right) \\ &= 3\alpha_{N_T}\log_3 N_V + \alpha_V \times 5V_{\text{step}}\left(\frac{N_V - 1}{2}\right) \end{aligned} \tag{2.46}$$

Figure 2.20 shows the ratio of these two cost functions versus $\frac{\alpha_{N_S}}{\alpha_{N_T}}$ for different N_V. This curve indicates that if $\frac{\alpha_V V_{\text{step}}}{\alpha_T} < X$, topology of Fig. 2.16 will be optimal and if

$\frac{\alpha_V V_{step}}{\alpha_T} > X$, the suggested topology in Fig. 2.17 will be optimal from C_4 criterion point of view. X depends on the value of N_V and according to Fig. 2.20, for $N_V = 25$, 50 and 50, X is 1, 0.6 and 0.5, respectively. It is noted that if even $\frac{\alpha_V V_{step}}{\alpha_T} > X$ is held, in the best conditions topology of Fig. 2.17 is 6% better than that topology of Fig. 2.16 from cost criterion C_4 point of view.

References

1. Dernirci O, Torrey DA, Degeneff RC, Schaeffer FK, Frazer R (1998) A new approach to solid-state on-load tap-changing transformers. IEEE Trans Power Deliv 13(3):952–961
2. Bauer P, deHaan SWH (1998) Electronic tap changer for 500 kVA/10 kV distribution transformers: design, experimental results and impact in distribution network. IEEE IAS 2:1530–1537

Chapter 3
Bi-Directional Solid-State Switches in Full-Electronic Tap-Changer

3.1 Introduction

The most important element in a full-electronic tap-changer is its solid-state switches. Since voltages and currents are AC, the switches used in electronic tap-changer must be bi-directional. Bi-directional switches are tolerable for the positive and negative voltages in the switch-off condition and passing the positive and negative currents in the switch-on conditions [1]. In this chapter realization of these switches is presented, static behavior and current commutation is explained.

3.2 Realization of Bi-Directional Switches

So far the solid-state device with full characteristic of a bi-directional switch has not been proposed; therefore a bi-directional switch must be realized by the combination of the solid-state switches. Figure 3.1 shows some samples of the possible cases to realize bi-directional switches.

In Fig. 3.1a, a bi-directional switch has been created by two anti-parallel thyristors. In the switch-off case, both thyristors are off, and the terminal voltage of the switch is tolerated by the thyristors in any direction. In the switch-on case, both thyristors are on and the current passes in both directions. Meanwhile, it is possible to control the current conduction direction. In this switch, there is no drop other than the thyristor drop in the switch-on condition, and this is the advantage of such a bi-directional switch because its losses in the on condition is identical with the losses in the switch-on condition of the thyristor. This arrangement for realization of a bi-directional switch is only possible by those switches that they can also stand the reverse blocking capability in the switch-off condition, so it is impossible to use switches such as MOSFET or IGBT.

Figure 3.1b presents realization of a bi-directional switch using a uni-directional switch (IGBT) and four diodes. In the switch-off condition, IGBT is off and

J. Faiz and B. Siahkolah, *Electronic Tap-changer for Distribution Transformers*, Power Systems, DOI: 10.1007/978-3-642-19911-0_3,

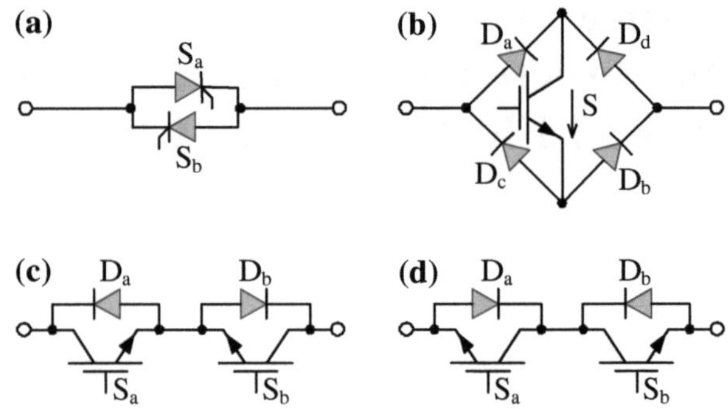

Fig. 3.1 Some samples of possible cases to realize bi-directional switches

the terminal voltage of the switch in any direction is applied to IGBT using four diodes (as positive). In the switch-on condition, IGBT will be on and the current in any direction will conduct in IGBT using four diodes (as positive). Positive directions of the voltage and current have been shown in Fig. 3.1b.

Since in this bi-directional switch only one solid-state switch has been used, its structure is simple and cost is low. On the other hand, the voltage drop of the switch in the switch-on condition is the sum of drop on IGBT switch and two diodes, therefore, the switch-on condition losses of this switch is large. In addition, the proposed switch has other drawback, it is impossible to control the direction of the current conduction; in other words, the current will conduct in any direction when IGBT is switch-on. In the proposed bi-directional switch, any other solid-state switch such as thyristor or MOSFET can be used instead of IGBT.

In Fig. 3.1c, a bi-directional switch has been created using two uni-directional switches (IGBT) and two diodes. In the switch-off condition, both IGBTs are off and the terminal voltage of the bi-directional switch is applied to S_a or S_b, depending on its direction. In the switch-on condition, also both IGBTs will be on, and the current passes through D_b, S_a of D_a, S_b depending on its direction. In this switch, voltage drop in the switch-on condition consists of the drop at the terminal of the switch IGBT and one diode. In this arrangement the current direction can be controlled by switching-on any of the S_a or S_b switches.

Arrangement of Fig. 3.1d is similar to that of Fig. 3.1c. The difference is that in Fig. 3.1c the emitters of two switches S_a and S_b are connected to each other, while in Fig. 3.1d the above-mentioned switches have common collectors. This difference does not cause a basic discrepancy in the performance of two switches; only the number of the required isolated supplies for driving the switches can vary in different topologies.

Table 3.1 compares the characteristics of the bi-directional switches of Fig. 3.1 briefly.

Since there is a need for bi-directional fast switches to be able to control the current direction, bi-directional switch (Fig. 3.1c) is used. The bi-directional switch mentioned in the next sections is the above-mentioned arrangement.

Table 3.1 Comparison of characteristics of the bi-directional switches of Fig. 3.1

Arrangement	No. of switches	No. of diodes	Switch-on condition drop	Capability of controlling current direction	Possibility of using fast switches such as IGBT and MOSFET
3.1a	2	0	Switch drop	+	– (only thyristor)
3.1b	1	4	Switch drop + drop of 4 diodes	–	+
3.1c	2	2	Switch drop + diode drop	+	+

3.3 Switch-On Condition Losses in Bi-Directional Switches

In this section, the switch-on condition losses of a bi-directional switch, shown in Fig. 3.2, are calculated. Suppose the on condition voltage drop of diodes D_a and D_b are equal to V_2, and the switch-on condition voltage drop of switches S_a and S_b are equal to $V_{CE_{sat}}$, then the terminal voltage of the bi-directional switch is as follows:

$$V_s = \begin{cases} V_\gamma + V_{CE_{sat}}, & I_s > 0 \\ 0, & I_s = 0 \\ -(V_\gamma + V_{CE_{sat}}), & I_s < 0 \end{cases} \tag{3.1}$$

So, the switch-on condition losses of the bi-directional switch for periodic current I_s is as follows:

$$P_{on} = \frac{1}{T}\int_T (V_\gamma + V_{CE_{sat}})|I_s|\mathrm{d}t \tag{3.2}$$

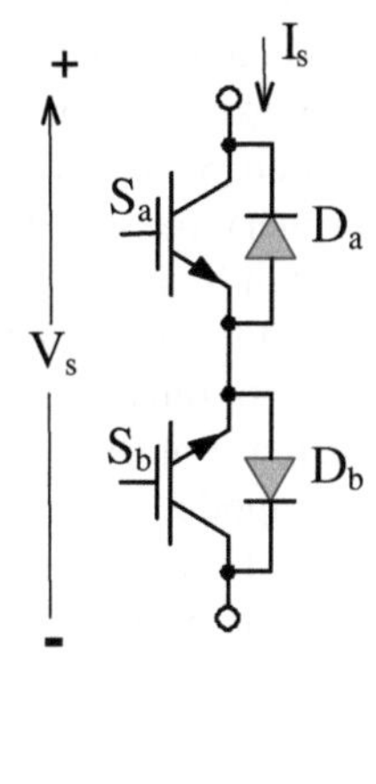

Fig. 3.2 A bi-directional switch used for calculation of on condition losses

where T is the period of current I_s. For the sinusoidal current of $\left(I_s = I_{SM}\sin\frac{2\pi}{T}t\right)$, Eq. 3.2 is simplified as follows:

$$P_{\text{on}} = \frac{1}{T}\int_0^T (V_\gamma + V_{\text{CE}_{\text{sat}}})|I_s|\text{d}t = \frac{2(V_\gamma + V_{\text{CE}_{\text{sat}}})}{T} I_{SM}\int_0^{\frac{T}{2}} \sin\left(\frac{2\pi}{T}t\right)\text{d}t$$
$$= \frac{2(V_\gamma + V_{\text{CE}_{\text{sat}}})}{T} I_{SM} \times \frac{T}{\pi} = \frac{2(V_\gamma + V_{\text{CE}_{\text{sat}}})I_{SM}}{\pi} \tag{3.3}$$

If the number of the bi-directional switches in the current path of the electronic tap-changer is shown by n (bi-directional switches are assumed identical) and the rated voltage of the transformer in the tap-changers side is shown by V_n, efficiency of the electronic tap-changer is as follows:

$$\eta = 1 - \frac{\frac{2(V_\gamma + V_{\text{CE}_{\text{sat}}})I_{\text{SM}}}{\pi} \times n}{V_n \frac{I_{\text{SM}}}{\sqrt{2}}} = 1 - 0.9 \times \frac{(V_\gamma + V_{\text{CE}_{\text{sat}}})}{V_n} \times n \tag{3.4}$$

It is noted that this efficiency has been obtained considering only the losses in the electronic tap-changer due to the drop of the switches for switch-on condition. If other losses such as winding and core losses of transformer as well as switching losses of the tap-changer switches are taken into account, efficiency will be lower. Suppose, there are six bi-directional switches in the current path of the electronic tap-changer, the related $V_{\text{CE}_{\text{sat}}} + V_\gamma$ is 4 V, and tap-changer switches are in the 20 kV side (rated voltage), based on Eq. 3.4, the efficiency will be 99.89% which is very desirable.

3.4 Current Commutation Between Bi-Directional Switches

Current commutation between the bi-directional switches is more difficult than that of the uni-directional switches in the base of the voltage source inverters; the reason is that there is not an idle path in the first one [1]. To explain this, a simple tap-changer of Fig. 3.3 is considered. In this figure bi-directional switches S_1 and S_2 make it possible to select tap number 1 and tap number 2 respectively. Suppose switch S_1 is on and switch S_2 is off; then, the load current transferred to the primary passes to switch S_1. Now consider the moment that switch S_1 is off and switch S_2 is on. At such a moment, if first switch S_1 is off and after a short dead-time the small switch "S_2" is on, both switches will be off in the dead-time period; this not only causes the disconnection of the load current path and probably large increase of the voltage, but in the switch-off period of both switches of S_1 and S_2, the applied voltage will be equivalent to the total line voltage (V_{in}); however, if one of these two switches is on, the applied voltage on them will be limited and there is no need to stand the total line voltage.

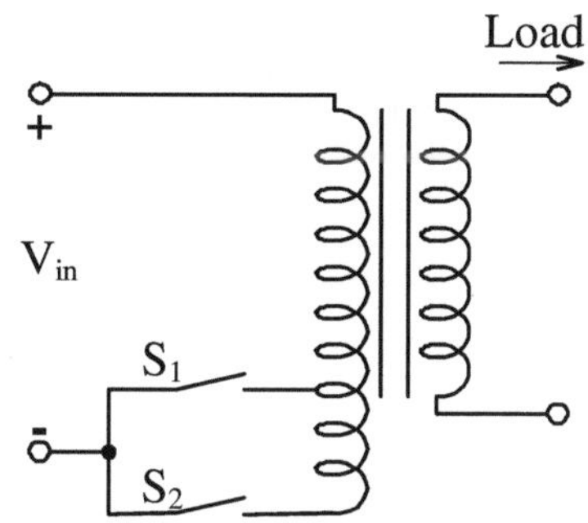

Fig. 3.3 Simple structure of tap-changer with two bi-directional switches S_1 and S_2 for explanation of requirements of switching during commutation between currents of two switches

To transfer from tap 1 to tap 2, first switch S_1 is on and after a short over-lap time, switch S_1 is off. During the over-lap time both switches will be on and this means the short circuit of the tap. This causes large short circuit currents and can damage the winding of the transformer or switches S_1 and S_2. Therefore, during the current commutation between switches S_1 and S_2 cannot put dead-time similar to the switches commutation in the voltage source inverter or have over-lap time similar for switches commutation in the current source inverter [2]. On the other hand, the switch-on or off time of a solid-state switch is not zero; in other words, the switch-on and off process is not instantaneous, so cannot at the same time switch-off one switch and on the other one. Therefore, the current commutation between the bi-directional switches needs paying a specific attention and control. So, the commutation requirements in the bi-directional switches can be summarized as follows:

(a) Both switches must not be off simultaneously; this means closing the current path and creating large voltages.
(b) Both switches must not be on; this means short circuiting of the taps and generating high circulating currents.

There are two basic methods to switch the switches for the correct current commutation between the bi-directional switches. The first method is based on the measurement of current direction, and the second method is based on the measurement of the terminal voltages of the switches. These two commutation methods are described in the next sections.

3.5 Commutation Based on Current Direction

In this method of commutation, the direction of passing current in the bi-directional switches is controlled. To explain this method a simple circuit of Fig. 3.4 is considered.

In this circuit, each bi-directional switch is connected to the voltage source and load is considered as a current source. If bi-directional switch S_1 is on, voltage

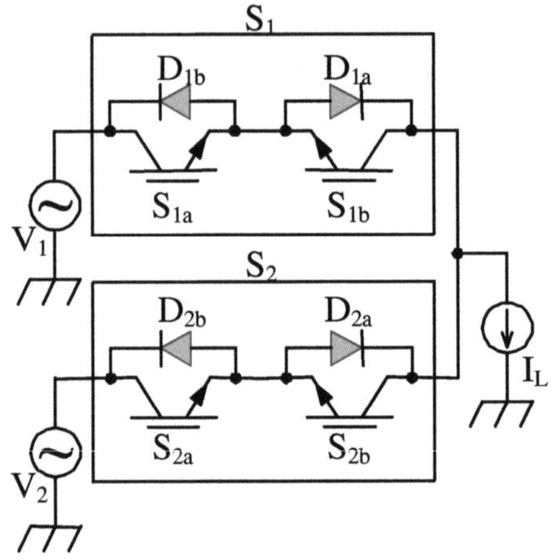

Fig. 3.4 Simple circuit to explain commutation based on current direction

source V_1, and if bi-directional S_2 is on, voltage source V_2 is applied to the load. To explain the commutation method based on the current, it is assumed that bi-directional switch S_1 is already on and decided to transfer the load current from that switch to the bi-directional switch S_2. Meanwhile, the load current at the commutation time is assumed in the same direction shown in Fig. 3.4. When bi-directional switch S_1 is on, it means that both S_{1a} and S_{1b} are switch-on to be able to pass the load current in both directions. The following stages are followed during commutation:

Stage 1. Direction of load current is detected by a suitable detector. Therefore, it is determined that which of two switches S_{1b} and S_{2b} actually carries the load current. If the load current direction corresponds with the direction shown in Fig. 3.4, S_{1a} will carry the current and if the load current direction is opposite to the above-mentioned, S_{1b} will carry the current. Since it is assumed that the load current corresponds to the direction shown in Fig. 3.4, switch S_{1a} carries the load current.

Stage 2. In bi-directional switch, S_1 is the switch that does not carry the load current (S_{1b}) and is off; this does not have any impact on the position of the current of switches.

Stage 3. In bi-directional switch S_2 is a switch that finally carries the load current (S_{2a}), and it is switch-on.

Stage 4. In bi-directional switch S_1, the switch carrying the current (S_{1a}) is off. Exchange of the load current from switch S_{1a} to S_{2a} is done in stage 3 or stage 4. This depends on the amplitudes of voltage sources V_1 and V_2 in the commutation time. If at the commutation time amplitude of voltage source V_1 is less than that of V_2, diode D_{1a} is inversely biased by switching on switch $S_{2a;}$ therefore, current will transfer from S_{1a} to S_{2a} (in stage 3), and if at the commutation time the amplitude of voltage source V_1 is higher than that of V_2, diode D_{2a} will be inversely biased by switching on switch S_{2a}. So, current will not be transferred to switch S_{2a} until switch S_{1a} is off (stage 4) [3].

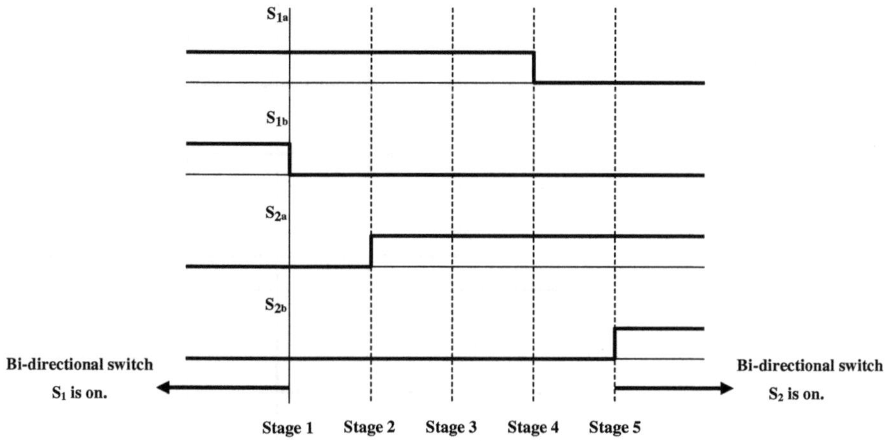

Fig. 3.5 Time diagram for commutation between two bi-directional switches based on current direction

Stage 5. Finally, switch S_{2b} will be on and bi-directional switch S_2 will be ready to pass the current in both directions. Time diagram of the mentioned stages have been shown in Fig. 3.5. Delay time between the different stages is specified considering the on and off times of the switches.

Of course, in the commutation method based on the current direction, (the current direction is always known) only the carrying current switches are on and off, in this case, stages 4 and 5 are eliminated from the mentioned stages and commutation is practically done in two stages.

If in the commutation method based on the current direction, the current direction is wrong or current direction changes during commutation stages, commutation will not be done correctly and large over-voltages and also damage of the switches are expected. To prevent this, some methods can be recommended [4].

3.6 Commutation Based on Voltage Amplitude

Contrary to the previous method, in commutation method based on the voltage amplitude, knowledge of the current direction is not necessary, but amplitudes of voltage sources V_1 and V_2 are important. The commutation stages are designed based on the larger amplitude of V_1 and V_2. First assume that the amplitude of voltage source V_1 is larger than that of V_2 at the commutation instant, and it is decided that bi-directional switch S_1 which was already on, is off and switch S_2 is on. To do this, the following stages are followed.

Stage 1. Switch S_{2a} is on. Since V_1 is larger than V_2, when switch S_{2a} is on, there will be no short circuit of the input voltage sources.

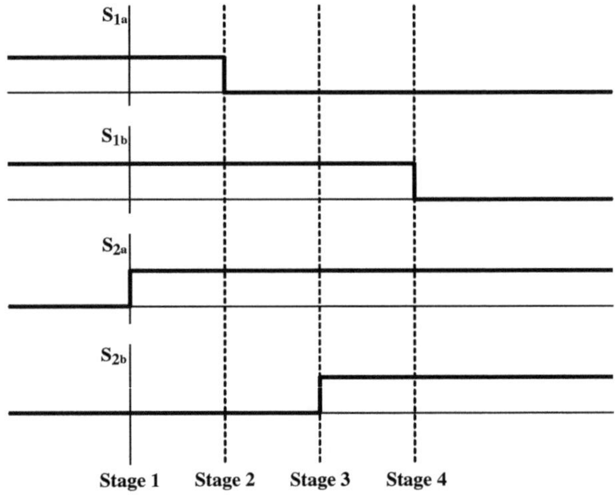

Fig. 3.6 Time diagram for current commutation from bi-directional switch S_1 to bi-directional switch S_2 based on amplitude of voltage in case of $V_1 > V_2$

Stage 2. Switch S_{1a} is off. If the direction of the current corresponds to that presented in Fig. 3.4, at this moment the current will transfer from switch S_{2a} to switch S_{2a}.

Stage 3. Switch S_{2b} is on. Since switch S_{1a} was off in the previous stage, when switch S_{2b} is on, there is the risk of short circuit. If direction of the current is opposite to the presented direction in Fig. 3.4, when switch S_{2b} is on, an inverse voltage will be applied to D_{1b}. Thus, the current will be transferred from S_{1b} to S_{2b}.

Stage 4. Switch S_{1b} will be on. Time diagram of the mentioned stages is shown in Fig. 3.6. Thus stages of the current commutation from bi-directional switch S_1 to bi-directional switch S_2 is ended, and it is noted that this method is not sensitive to the direction of the load current.

Stages of current commutation from bi-directional switch S_1 to bi-directional switch S_2 when $V_1 < V_2$ are briefly as follows:

Stage 1. Switch S_{2b} is on.
Stage 2. Switch S_{2b} is off.
Stage 3. Switch S_{2a} is on.
Stage 4. Switch S_{1a} is off.

Time diagram of the above stages have been shown in Fig. 3.7.

It is noted that contrary to the commutation method based on the current direction, in this commutation method if the current direction changes in the commutation stages, there will be no risk because the method is not basically sensitive to the current direction and there is always passing path for current in both directions. This is clearer referring to time diagram of 3.6 and 3.7. It is seen that in any stage of the commutation stages, at least one of two switches S_{1a} or S_{2a}, and also at least one of switch S_{1b} or S_{2b} is on. This means that there is a path for the current in both directions.

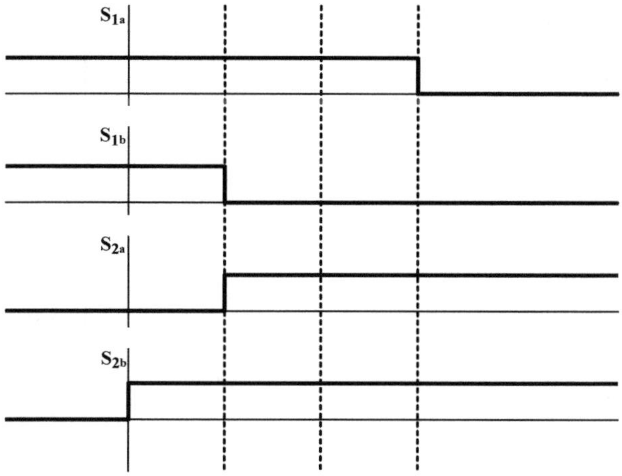

Fig. 3.7 Time diagram for current commutation from bi-directional switch S_1 to bi-directional switch S_2 based on amplitude of voltage in case of $V_1 < V_2$

Table 3.2 Comparison of two commutation methods based on voltage and current direction

No.	Commutation method	Current sensor	Sensitive to current direction during commutation	Short circuit risk due to measurement error	Require stages in commutation
1	Current direction	Exist	Yes	Exist	4 stages can be reduced to 2
2	Voltage amplitude	Non-exist	No	Non-exist	4 stages

Of course, if measured voltages V_1 and V_2 are wrong, or detection of larger voltage is incorrect or sign of voltage $V_1 - V_2$ changes during commutation stages, short circuit will be possible. Since these errors or sign changing occur in small voltages and basically in those voltages, voltage difference of taps is even less than that of the conduction threshold of the switches, no short circuit current is created.

Table 3.2 compares the specifications of the two commutation methods briefly. It is seen that commutation method based on the voltage is more safe and simpler compared to the commutation method based on the current direction.

3.7 Commutation of Current Between Bi-Directional Switches in Electronic Tap-Changer

So far, current commutation between the bi-directional switches has been based on a simple circuit of Fig. 3.4. The situation differs in an electronic tap-changer due to existing leakage inductance [5]. Figure 3.8 shows the equivalent circuit of tapped-winding of a tap-changer with two taps.

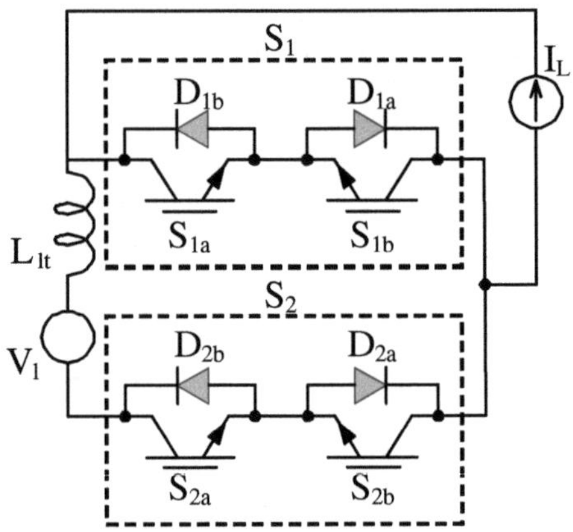

Fig. 3.8 Equivalent circuit of tapped-winding of a tap-changer with two taps

Current commutation between switches S_1 and S_2 have more losses and over-voltage due to tap winding leakage inductance L_{1t}. To clarify this, the current commutation stages based on the current direction is described for circuit presented in Fig. 3.8. Suppose that the direction of the load current is as shown in Fig. 3.8 and its value V_1 is positive at the commutation instant. The following stages are for commutation.

Stage 1. Switch S_{1b} is off. This does not change the current pass, and the current still passes through switch S_{1a} and diode D_{1a}.

Stage 2. Switch S_{2a} is on. In this stage V_1 is positive and applied voltage on diode D_{2a} will be inverted and thus the load current from switch S_{2a} will not be transferred.

Stage 3. Switch S_{1a} is off. Zero gate command of switch S_{1a} diminishes the current of switch S_{1a} with slope α (which depends on the specifications of the switch). In this case, the remaining current of I_L passes through switch S_{2a}, diode D_{2a} and inductance L_{1t}. So, switch-on voltage of S_{1a} at the off-time is obtained as follows:

$$V_{S_{1a}} = V_1 + L_{1_t}\alpha \tag{3.5}$$

It is seen that voltage of switch S_{1a} at the off-time is larger than V_1 by $L_{1t}\alpha$, and this seriously increases the losses of switch S_{1a} at its off-time. It also may damage switch S_{1a}. Figure 3.9 shows the voltage and current of the circuit.

Stage 4. Switch S_{2b} is on. Thus, bi-directional switch S_2 is ready to pass the current in both directions.

Certainly, larger value of L_{1t} will generate larger over-voltages. The simplest method to solve this problem is the use of a snubber capacitor at the terminals of tap (Fig. 3.10). This capacitor does not allow creating large over-voltage. Selection procedure for this capacitor is presented in Sect. 3.8.

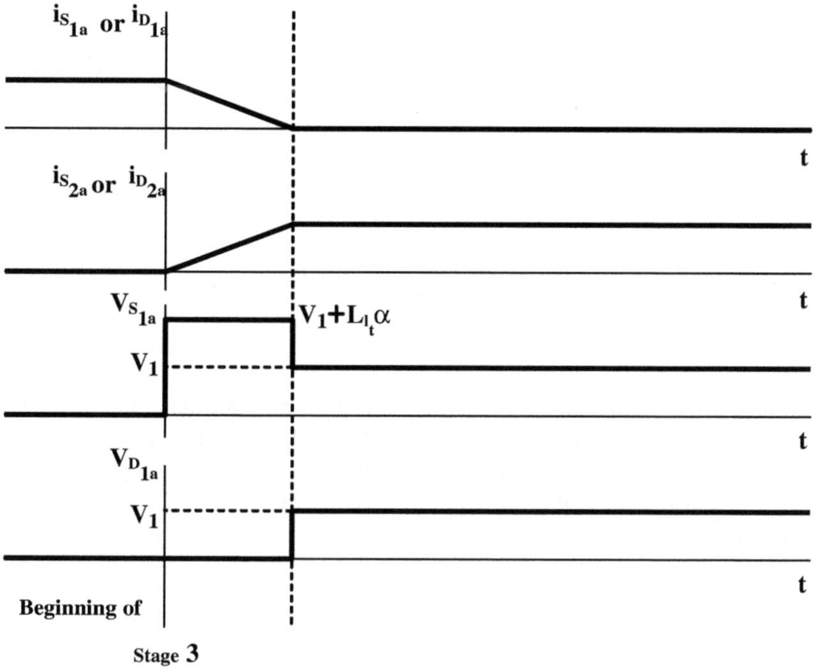

Fig. 3.9 Voltage and current waveforms of switches during stage 3 from commutation based of current direction assuming $V_1 > 0$

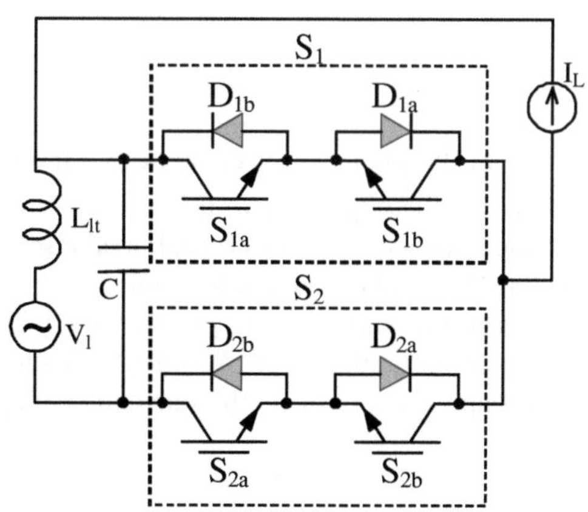

Fig. 3.10 Snubber capacitor placement for compensating over-voltages due to leakage inductance L_{1t}

3.8 Design of Snubber Capacitor at Terminal of Electronic Tap-Changer

Snubber capacitor C can be placed between the two taps and solves the problem of large over-voltages due to the leakage inductance L_{1t}. In this section, performance of this capacitor is more precisely studied and required equations for designing this

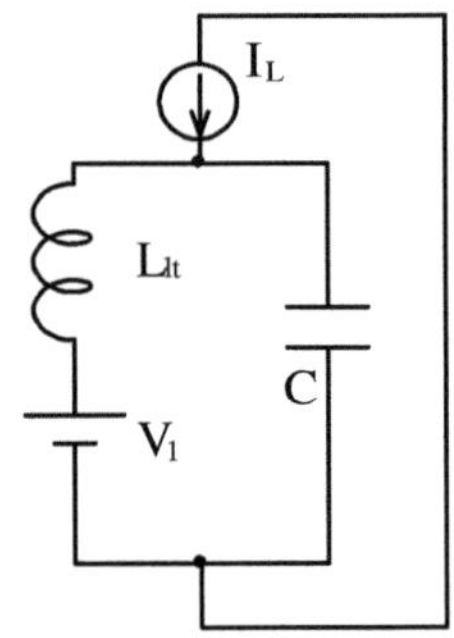

Fig. 3.11 Equivalent circuit of Fig. 3.10 during commutation from switch S_1 to switch S_2

capacitance are introduced. It is assumed that the time constants due to the circuit elements are very longer than that of the off-time of switch S_{1a}, but is very shorter than that of the industrial electricity period. By this assumption, equivalent circuit of Fig. 3.10 during commutation can be replaced by Fig. 3.11. The initial conditions in this circuit are as follows:

$$\begin{aligned} I_{1t}(0) &= 0 \\ V_C(0) &= V_1 \end{aligned} \tag{3.6}$$

If in equivalent circuit of Fig. 3.11, the parasitic resistors and switches drops are neglected, it can be easily shown that the voltage and snubber capacitance follows the following relationship:

$$V_C = V_1 + \sqrt{\frac{L_{l_t}}{C}}\, I_L \tag{3.7}$$

So, in the presence of capacitor C, the over-voltage of $\sqrt{\frac{L_{l_t}}{C}}\, I_L$ is generated. By increasing the capacitance, this over-voltage is limited to the proposed value; for instance, for $L_{1t} = 200\ \mu H$, peak load current 40 A, maximum permissible voltage of 200 V, snubber capacitor C must be 8 μF. By this capacitance, and assuming 500 ns commutation stages period, total load current passing through this capacitor during commutation is 40 A, only 2.5 V charges the capacitor during commutation period and this confirms the constant voltage of capacitor C during commutation period.

3.9 Switching Losses in Bi-Directional Switches

Switching losses of switches during the commutation stages are analyzed in this section. The time of starting the current commutation from switch S_1 to switch S_2 in Fig. 3.4 is summarized as follows:

(I) If $I_L > 0$ {If $V_1 - V_2 < 0$ then at the on-time of switch S_{2a} commutaion occurs (stage 3).
{If $V_1 - V_2 > 0$ then at off-time of switch S_{1a} commutation occurs (stage 4).

(II) If $I_L < 0$ {If $V_1 - V_2 < 0$ then at the off-time of switch S_{1b} commutaion occurs (stage 4).
{If $V_1 - V_2 > 0$ then at on-time of switch S_{2b} commutation occurs (stage 3).

The losses of each switches can be evaluated in different cases. Here, only the losses of switches are considered due to their similarity.

Figure 3.12 shows the voltage and current of each switch during commutation period for $V_1 - V_2 < 0$. As seen in these conditions, commutation is carried out practically between diode D_{1a} and S_{2a}. In this case, switch on of S_{2a} has slight losses due to the short period of the off-time of diode D_{10}. However, switch S_{1a} is off in ZCS and lossless conditions.

Figure 3.13, shows the voltage and current of each switch during commutation period for $V_1 - V_2 > 0$. In this case, the switch off of S_{1a} has considerable losses, but switch S_{2a} is in lossless conditions of ZCS.

Comparison of waveforms 3.12 and 3.13 indicates that the switching losses in the case of $V_1 - V_2 < 0$ (for $I_L > 0$) is less than the case $V_1 - V_2 > 0$. The reason is that in the first case commutation is done between diode D_{1a} and switch S_{2a}, and it has slight losses due to short period of this process; but in the second case, commutation is between S_{1a} and S_{2a} and in fact during off-time of switch S_{1a}, this switch has to suffer voltage $V_1 - V_2$, thus losses in this case is larger than the

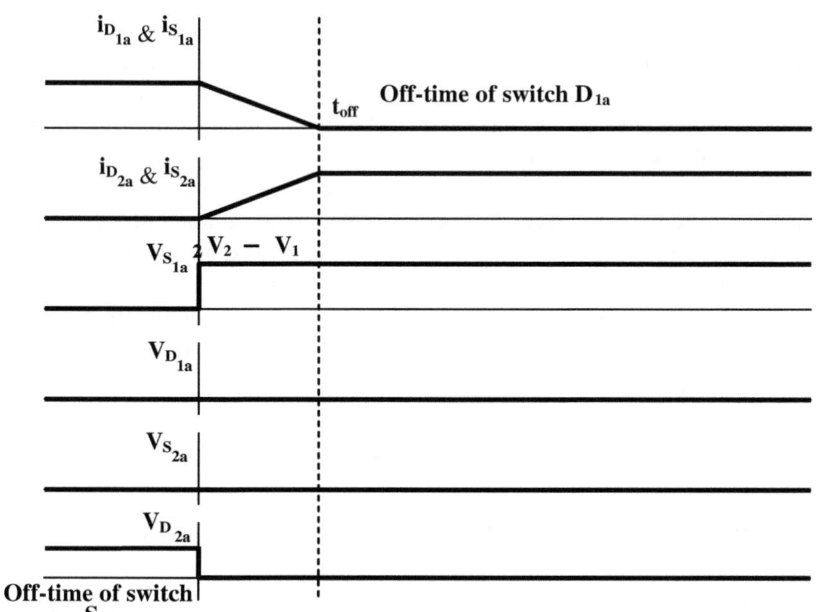

Fig. 3.12 Voltage and current waveforms of switches during commutation period for $V_1 - V_2 < 0$

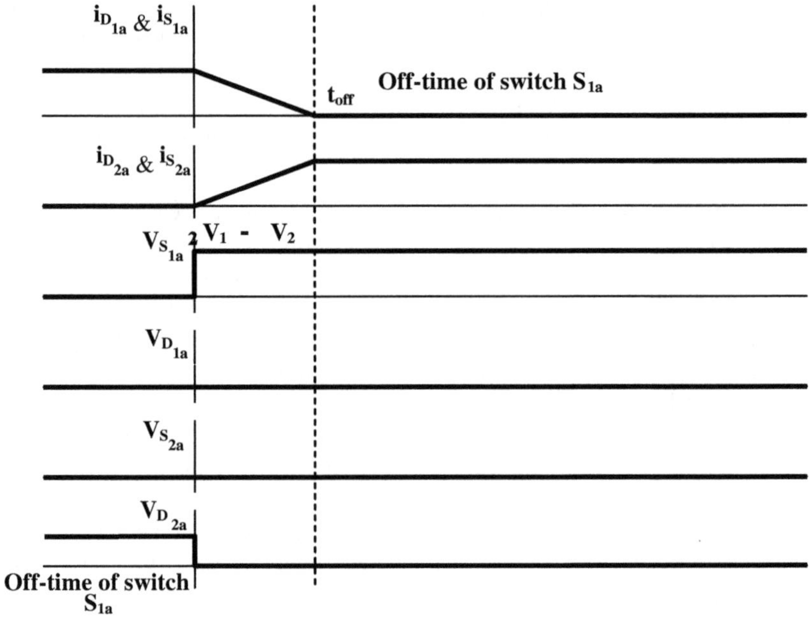

Fig. 3.13 Voltage and current waveforms of switches during commutation period for $V_1 - V_2 > 0$

previous case. Since both the first and second cases takes 50% of period, this commutation method is called semi-soft current commutation [4]. Of course, realization of the soft switching methods have been also suggested for switching bi-directional switches [6–8], but these techniques lead to a highly complicated circuit and only is used in special applications.

If in electronic tap-changer the permissible instants of switches position change is limited to the zero current crossing instants, not only the switching losses tend to zero, but over-voltages due to the leakage inductances will be largely reduced, so snubber capacitors will be required. The only problem with method is the limitation of the permissible instants of the changing the switches position that can create maximum a half cycle of the industrial power system delay in the tap-changer operation. If PWM is not used, this situation is fully acceptable [9, 10].

3.10 Modulation of Switches in Electronic Tap-Changer

There are four modulation methods for modulation of electronic tap-changer switches for controlling output voltage. These methods are described in the following parts.

3.10.1 Phase Control Modulation

As shown in Fig. 3.14, in this type of modulation, over a part of a cycle, a tap is connected to the output and in the remaining cycle other tap is connected, and by controlling each tap (time α in Fig. 3.14) the rms value of the output voltage can be controlled. Although in this method the output voltage can be continuously controlled by adjusting α, the basic drawback of this method is generating low frequency harmonics in the output that is undesirable in most cases.

3.10.2 Discredited Cycle Modulation

In this type of modulation, over one repetitive cycle equivalent with N cycles of electrical supply, M cycles of a tap is connected to the output, and N-M remaining cycles of the tap is connected. In this modulation method, the rms value of the output voltage can be adjusted. Figure 3.15 shows a typical output waveform of this modulation method. This method is unable to provide continuous output voltage and produces flicker. The advantage of this method is that without adding tap in transformer, it is possible to generate rms voltages between the taps voltage.

3.10.3 Pulse Width Modulation

In this type of modulation, the attempt is made to adjust and control the output voltage instantaneously by changing the taps in high frequency. A typical output waveform using this modulation method has been shown in Fig. 3.16. This method

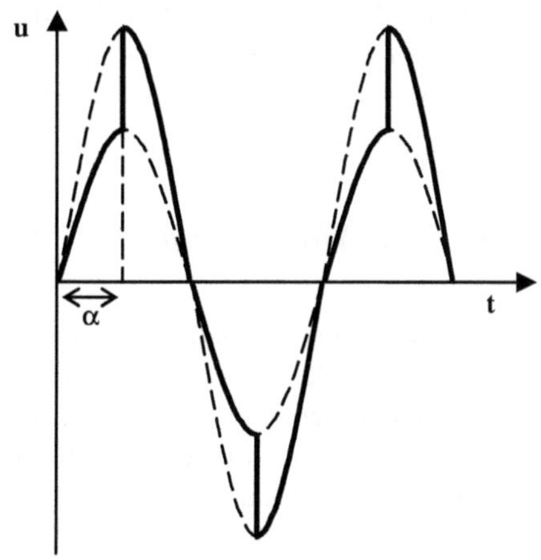

Fig. 3.14 Output waveform in phase control modulation method

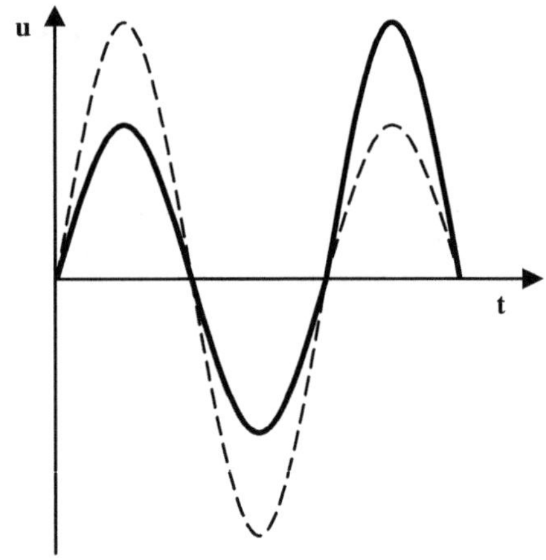

Fig. 3.15 Output waveform in discredited cycle modulation method

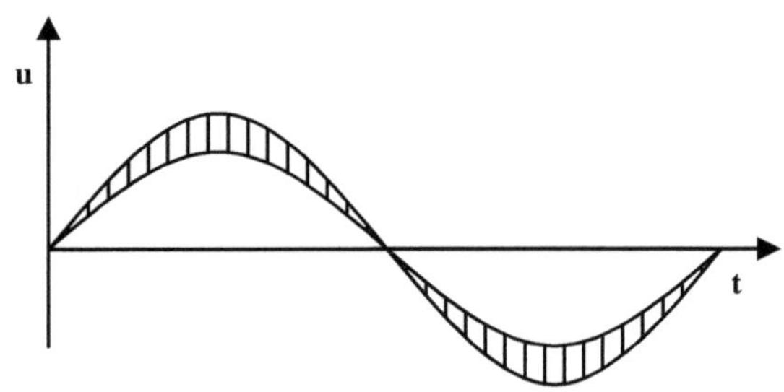

Fig. 3.16 Output waveform in pulse width modulation method

of modulation, is certainly very desirable because of continuous adjustment (possibility of the modification of waveform). However, the use of this type of modulation causes high frequency harmonics in the output. To eliminate these harmonics, some filters must be inserted in the output. Meanwhile, implementation of this modulation method will produce large switching losses due to the high switching frequency, so application of soft switching technique will be necessary which increases the complexity of the system.

3.10.4 No Modulation

In order to adjust the output voltage in third case, the required tap is selected and connected to the output by tap-changer and until the output voltage is in permissible range, the taps position does not change. If due to any reason the output voltage leaves its permissible level, tap-changer changes the taps position, and an appropriate tap is connected to the output. Certainly, no extra harmonic is added permanently to the output. Operation of the electronic tap-changer in this

Table 3.3 Comparison of different modulation methods in electronic tap-changer

No.	Modulation method	Extra harmonic in output	Voltage adjustment	Modification of voltage waveform	Switching losses
1	Phase control	Low frequency	Continuous	No	Average
2	Discredited pulse	Low frequency	Discredit	No	Average
3	Pulse width	High frequency	Continuous	Yes	High
4	No modulation	No harmonic	Discredit	No	Low

method is similar to the mechanical tap-changers. In this book the mentioned method is used to adjust the output voltage. Table 3.3 compares specifications of the four above-mentioned methods.

References

1. Wheeler P, Clare J, Empringham L, Apap M, Bland M (2002) Matrix converters. Power Eng J 273–282
2. Bauer P, Schoevaars R (2003) Bidirectional switch for a solid-state tap-changer. IEEE PESC 03, pp 466–471
3. Wheeler PW, Grant DA (1992) Reducing the semiconductor losses in a matrix converter, IEE colloquium on variable speed drives and motion control, pp 14/1–14/5
4. Empringham L, Wheeler PW, Clare JC (1998) Matrix converter Bi-directional switch commutation using intelligent gate drives, IEE conference on power electronics and variable speed drives, Sept 1998, pp 626–631
5. Y-Zai FQ, O'Kelly D (1996) Solid-state on-load transformer tap-changer. IEE Proc Electric Power Appl 143(6):481–491
6. Teichmann R, Oyama J (2002) ARCP soft-switching technique in matrix converters. IEEE Trans Ind Electron 49(2):353–361
7. Pan C, Chen T, Shieh J (1993) A zero switching loss matrix converter, IEEE PESC 93, pp 545–550
8. Bland M, Empringham L, Clare J, Wheeler P (2002) A new resonant soft switching topology for direct AC-AC converters, PEMC PESC, Sept 2002, pp 72–77
9. Faiz J, Siahkolah B (2007) New controller for an electronic tap-changer Part I: design procedure and simulation results. IEEE Trans Power Deliv 22(1):223–229
10. Faiz J, Siahkolah B (2007) New controller for an electronic tap-changer Part II: measurement algorithm and test results. IEEE Trans Power Deliv 22(1):230–237

Chapter 4
Design of Controller of Full-Electronic Tap-Changer for Output Voltage Regulation

4.1 Introduction

In most cases, the main goal for application of tap-changer system is adjusting and regulating the amplitude of output voltage. In these cases, the main responsibility of controller in the tap-changer system is minimization of the voltage amplitude deviation (rising and falling of voltage) in respect to the reference voltage at a point technically called regulation bus. This bus can be far from the secondary of transformer. In other words, the main aim of the controller is keeping the voltage amplitude at regulation bus in a defined range [1].

In addition to the above-mentioned main duty, in some events such as over load, output short circuit, input transient, over-voltages and fault of tap-changer, it is the duty of a tap-changer system to control those events and to adopt necessary decision for protecting the tap-changer system and also its connected loads.

The responsibilities that mentioned above for the tap-changer control system, are true for any tap-changer systems concerning all mechanical and electronic versions. However, taking into account the existing basic differences between the capabilities and also characteristics of the mechanical tap-changers and electronic tap-changers, there must be also some differences between their controllers. If the conventional controllers of mechanical tap-changers are used in the electronic tap-changer, many of capabilities and characteristics of the electronic tap-changer will not be realized.

By quick review of the differences in specifications of tap changing process in the mechanical tap-changers and electronic tap-changers, the vital differences in their controllers will be proposed in Sect. 4.2. Conventional controller used in the mechanical tap-changers is reviewed in Sect. 4.3. To do this, first the elements of the mechanical tap-changer system is modeled. Then, the relevant controller is propounded. By applying some modifications in the conventional controller of the mechanical tap-changers, the general structure of the electronic tap-changer

J. Faiz and B. Siahkolah, *Electronic Tap-changer for Distribution Transformers*, Power Systems, DOI: 10.1007/978-3-642-19911-0_4,

controller is obtained in Sects. 4.4 and 4.5. Section 4.6 presents the simulation results for electronic tap-changer with modified controller. Then, in Sect. 4.7, the electronic tap-changer will be systematically designed with completely novel glance. Simulation results will show the accuracy of every stage of the design. Finally, methods of the fast measurement of the voltage amplitude are discussed and compared, and finally the most appropriate method is selected.

4.2 Differences of Controller in Mechanical and Electronic Tap-Changers

Generally, tap changing process in the mechanical tap-changer is slow, non-continuous and step-by-step process. Non-continuous tap-changing process is a fundamental characteristic, and it originates from non-continuity nature of the voltage regulation using transformer taps changing and it is also true for the full-electronic tap-changers (of course, if high frequency modulation techniques are not used). Mechanical nature of the mechanical tap-changers results in a slow and step-by-step tap-changing process which is not the case in the electronic tap-changers. These differences in the characteristics of the mechanical and electronic tap-changers cause the following differences in the goals and responsibilities of the controllers of these two systems [2]:

(a) There is restriction in the numbers of permissible tap-changing in mechanical type, so reduction of the taps changing frequency is one of the controller aims, however, basically there is no such restriction in electronic tap-changer, and reduction of the taps changing frequency is not a part of the controller goals.
(b) In mechanical type, only tap changing from one position to its previous or next position is possible (step-by-step process). This is considered one of the controller limitations in these systems while there is no such limitation in electronic tap-changer, and it is possible to jump from the lowest tap to the topest tap, in any tap changing time.
(c) Essentially, tap changing process is slow in mechanical tap-changer, and there is no need to use advanced and fast detection algorithms to detect the rising and falling of the regulation bus voltage; averaging methods can be utilized over several cycles. However, in electronic tap-changers fast detection of voltage changes in the regulation bus is essential due to very fast tap changing; therefore, averaging method as slow methods cannot be applied. More explanations on the advanced voltage detection in the regulation bus will be presented in Sect. 4.8.
(d) In electronic tap-changer a memory is required to store the position of each switch for any output position due to the complicated taps and also tap-changer switches; however, there is no such need in the mechanical tap-changer.

4.3 Conventional Controller in Mechanical Tap-Changers

Figure 4.1 shows the block diagram of the mechanical tap-changer. This system consists of the following parts [1].

(a) Tapped-transformer.
(b) Motor drive unit and tap-changing mechanism.
(c) Current transformer (CT) and potential transformer (PT).
(d) Central control unit.
(e) Protection circuits.

The above-mentioned parts will be modeled in the following section.

Tapped-transformer voltage in the sinusoidal steady-state is as follows:

$$V_2 = \frac{V_1}{a} - Z_T(a)I_2 \tag{4.1}$$

where V_1, V_2 and I_2 are the primary voltage phasor, secondary voltage phasor and secondary (load) current phasor respectively. The ratio of the primary to the secondary turns of the transformer, a, is changed by changing the tap position. $Z_T(a)$ is the sum of secondary referred series impedances of transformer which is naturally depends on a. Taps changing, changes a. If the taps are placed on the primary of the transformer and designed linearly, changing a with tap-changing is represented as follows:

$$a = a_0 + n\Delta a \tag{4.2}$$

where a_0 is the ratio of the primary to secondary turns. Δa is the change of a in one step and n is the number of tap position (n varies from $-n_{\text{Mac}}$ to $+n_{\text{Max}}$). The number of tap position changed as follows:

$$n_{(\text{i})} = n_{(i-1)} + b_{(i)} \tag{4.3}$$

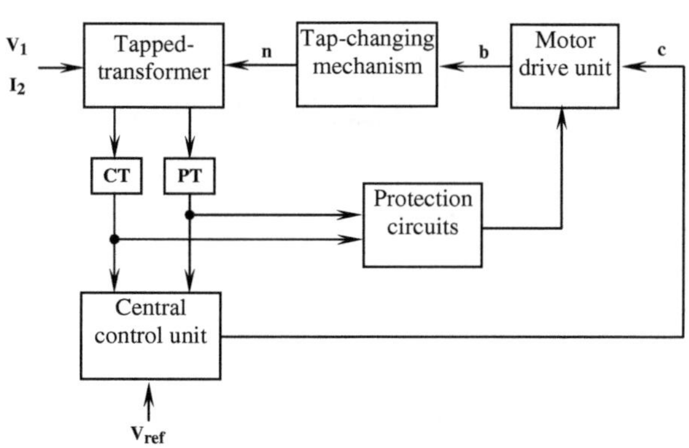

Fig. 4.1 Block diagram of a mechanical tap-changer

where b is the output signal of the motor drive unit and is -1 or 0 or 1. Index i means the present position and index $(i - 1)$ is the previous position. Signal b is in fact the output signal of the central control unit (c) which is applied to the tap changing mechanism after a time delay (T_m) due to the motor drive unit. Thus, b can be written as follows:

$$b(t) = c(t, T_m) \tag{4.4}$$

If regulating bus is far from the secondary of transformer, the transmission line drop is modeled as follows:

$$V_{2m} = V_2 - I_2 Z_C \tag{4.5}$$

where V_{2m} is the regulation bus voltage phasor, Z_C is the transmission line impedance between the secondary of transformer and regulation bus. Amplitude of voltage V_{2m} is compared with the reference voltage (V_{2r}) and the difference (ΔV) provides the main data for central control unit to determine c:

$$\Delta V = V_{2r} - |V_{2m}| \tag{4.6}$$

Figure 4.2 indicates that the conventional controller system is a bang-bang controller and selects c based on the value of ΔV, and applied to the tap-changer system after a T_d delay [3, 4]. T_d delay is not a constant delay, but it is decreased by increasing ΔV. Figure 4.3 demonstrates the variations of T_d versus ΔV. In Fig. 4.3, vertical and horizontal axes have been normalized versus DB and T_{d0} (constant values) respectively. DB is a dead-band (DB) distance and if ΔV is less than this value, tap-changing command will not be issued. DB is determined based on the value of Δa. If this dead-band is taken too small, the closed-loop control circuit will be instable, and if it is selected too large, accuracy of the bus voltage regulation will be decreased. T_{d0} also is selected based on the tap-changer system delay and delay due to the measurement of $|V_{2m}|$. If T_{d0} is not precisely set, the control loop oscillates and/or the control system becomes too slow.

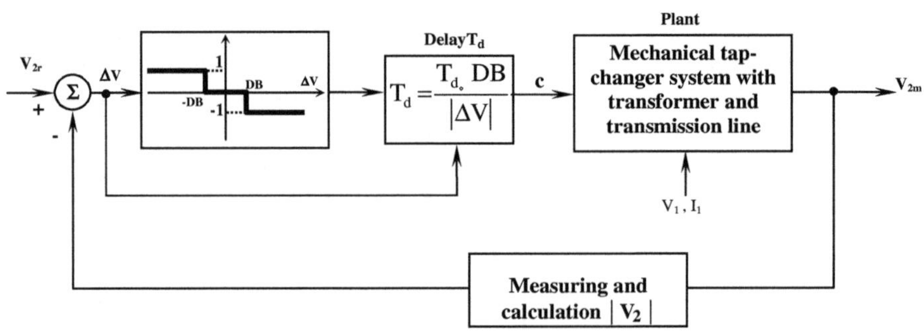

Fig. 4.2 Conventional controller of mechanical tap-changer for voltage regulation in regulation bus

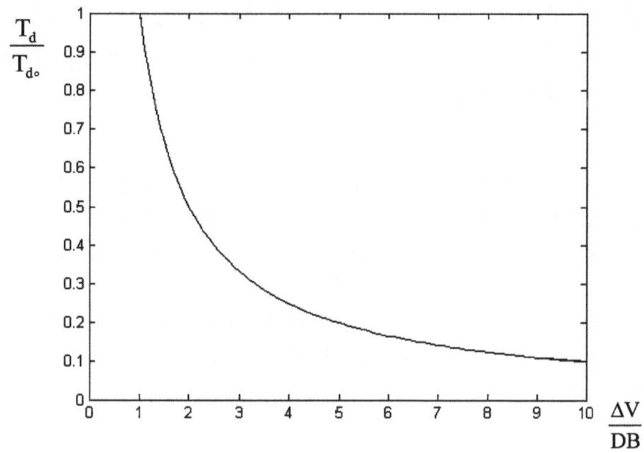

Fig. 4.3 Variations of T_d versus ΔV

4.4 Modification of Conventional Controller for Electronic Tap-Changer

A straightforward method for designing a control system in electronic tap-changer is that the conventional controller used in the mechanical tap-changers is used as base; then, this controller is modified based on the requirements of the electronic and mechanical tap-changers. Figure 4.4 presents such modified controller.

The applied modifications in this controller are as follows:

(a) Changing block of measuring and calculating the rms value to the block of measuring and calculating rms instantaneous value.
(b) Changing DB and saturation block to DB block.
(c) Changing delay block to compensating block (generally compensator may not be a simple delay).
(d) Adding a quantizer block, sampling and storing block, detecting zero crossing block and look-up table.

The above-mentioned items are described in the following section.

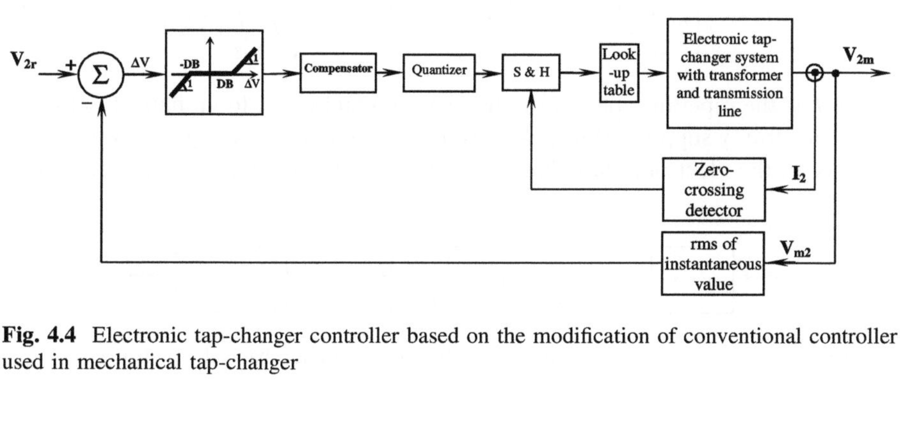

Fig. 4.4 Electronic tap-changer controller based on the modification of conventional controller used in mechanical tap-changer

4.4.1 Changing Block of Measuring and Calculating the rms Value to the Block of Measuring and Calculating rms Instantaneous Value

Figure 4.5 is used to measure the rms voltage of the regulation bus in a mechanical tap-changer. First signal $V_{2m}(t)$ is multiplied to itself, then, it is passed through a low-pass filter. This low-pass filter is designed as such that the harmonic with frequency of 100 Hz and higher in signal $V_{2m}^2(t)$ are deleted and its DC component is extracted. To do this, 3 dB frequency of the mentioned filter must be selected smaller than 0.1 times of 100 Hz. In other words, the following relationship must be held:

$$\frac{a}{2\pi} \leq 0/1 \times 100 = 10 \tag{4.7}$$

Since the proposed low-pass filter generates a delay around 4 times of the time constant for DC component, if (4.7) is satisfied, the following relationship is deduced:

$$\begin{aligned}\text{Delay of low-pass filter for DC component} &= 4 \times \text{filter time constant} \\ &= 4 \times \frac{2\pi}{a} \geq 0.4s\end{aligned} \tag{4.8}$$

It means that the minimum low-pass filter delay is equal to 0.4 s or 20 cycle of the public electricity supply frequency. Since tap-changing in the mechanical tap-changer is a slow process, this delay cannot create problem in mechanical tap-changer system, but there need on-time and fast measurement of rms value of the regulation bus voltage in the electronic tap-changer because tap-changing in this tap-changer is very fast. Therefore, concept of averaging rms value of the voltage cannot be used. One of the best substitutions for the rms value of voltage is rms of instantaneous voltage value. Figure 4.6 presents the block diagram of this method.

Considering Fig. 4.6, rms of instantaneous voltage value is as follows:

$$V_{2m_{rms}}(t) = \sqrt{\frac{1}{T}\int_0^t \left[V_{2m}^2(t') - V_{2m}^2(t'-T)\right]dt'} = \sqrt{\frac{1}{T}\int_{t-T}^t V_{2m}^2(t')dt'} \tag{4.9}$$

where T is the repetition period of signal $V_{2m}^2(t)$ and equal to an half of period of public electricity supply (10 ms). Rms of the instantaneous voltage is a very useful quantity in the electronic tap-changer. This quantity contains the data of an half

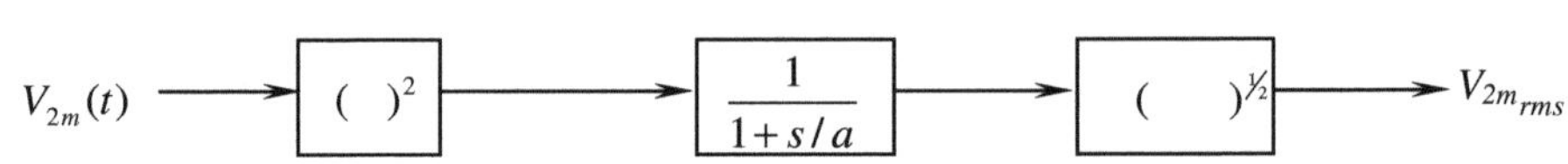

Fig. 4.5 Block diagram of rms voltage of regulation bus in a mechanical tap-changer

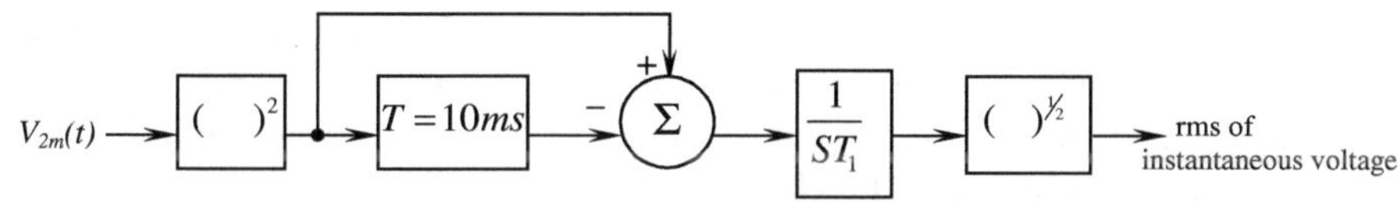

Fig. 4.6 Block diagram of measuring system of instantaneous voltage of regulation bus in an electronic tap-changer

cycle of the waveform and it measures the variations of the amplitude of output voltage in maximum of 10 ms.

4.4.2 Changing DB and Saturation Block to DB Block

There is a dead-band after calculation of ΔV due to the inherent limitation of the tap-changer and non-continuity of the system, and this dead-band is always necessary. In the mechanical tap-changer, it is only allowed to change a tap to one higher or lower tap, and ΔV goes to saturation block with values of -1 and $+1$ after passing dead-band block. However, in the electronic tap-changer it is possible to jump between the taps and there is basically no such saturation block.

4.4.3 Changing Delay Block to Compensating Block

On contrary to the mechanical tap-changer, the electronic tap-changers are fast system. Therefore, the compensator used in the electronic tap-changer must be designed more precisely such that it satisfies the control loop requirements and also does not slow down the system. Thus, in this system a general delay block is used in place of the delay; the latter is a particular type of compensator. The stages of compensator block design will be discussed in Sect. 4.6.

4.4.4 Adding a Quantizer Block, Sampling and Storing Block, Detecting Zero Crossing Block and Look-Up Table

Since the tap order number is an integer, quantizer block must be added. The range of the output quantizer block is between the maximum ($+n_{\max}$) and minimum ($-n_{\max}$) tap order number. The sampling and storing blocks and also zero-crossing detector are used in order to allow the tap-changing at the zero-crossing instants.

The reason for the existing lookup table block is that any row of the tap number corresponds with a particular position of the switches of the electronic tap-changer, and these positions of switches have been already stored in the memories of the lookup table block for further use.

4.5 Selection of Parameters in Modified Controller

Procedure for selecting the controller parameters of Fig. 4.4 is proposed in the following sub-sections.

4.5.1 Selection of DB

Tap-changer system is basically a non-continuous system for controlling the output voltage; thus, such a system cannot necessarily tends ΔV error to zero, and this the reason to have DB. So in order to select DB, the maximum error of ΔV must be calculated. To do so, first regulation steps V_2 must be calculated. Regulation step V_2 ($V_{2\text{step}}$) obtained by changing an order in tap number is given as follows:

$$V_{2\text{step}} = \left| \frac{V_1}{a_0 + n_1 \Delta a} - \frac{V_1}{a_0 + (n_1 + 1)\Delta a} \right| \cong \frac{V_1}{a_0} \times \frac{\Delta a / a_0}{1 + (2n_1 + 1)\Delta a / a_0} \tag{4.10}$$

where the change of I_2 as a result of changing a number in tap has been ignored. In Eq. 4.10, n_1 is the tap order number ($-\text{n}_{\max} \leq \text{n}_1 \leq n_{\max}$). The peak value of $V_{2\text{step}}$ is obtained for maximum value of V_1 and minimum n_1. Thus, the peak value of $V_{2\text{step}}$ is as follows:

$$V_{2\text{step}_{\max}} = \frac{V_{1\max}}{a_0} \times \frac{\frac{\Delta a}{a_0}}{1 + (-2n_{\max} + 1)\frac{\Delta a}{a_0}} \tag{4.11}$$

If the controller system is designed as such that DB be a constant parameter, then DB must be selected to satisfy the following inequality:

$$2DB > V_{2\text{step}_{\max}} \tag{4.12}$$

If parameter DB is taken to be a variable, then depending on the values of V_1 and n_1, it can be calculated satisfying the following inequality:

$$2DB > V_{2\text{step}} = \left| \frac{V_1}{a_0 + \Delta a(n_1)} - \frac{V_1}{a_0 + \Delta a(n_1 + 1)} \right| \tag{4.13}$$

It is noted that if i in determination of DB inequality (4.12) or (4.13) is not followed, the control loop will be oscillatory.

4.5.2 Selection of Parameters of Quantizer Block

Figure 4.7 shows the input–output characteristic of the quantizer block. In fact output of the block is the tap order number. Non-continuity of the system is clear in Fig. 4.7.

4.5.3 Selection of Compensator Block

In the electronic tap-changer, it is necessary to design compensator block more precisely in order to compensate the output voltage deviations in minimum possible time. Meanwhile the system must not be unstable in any case. It is interesting qualifiedly to use an integrator in the compensator block of the electronic tap-changer. The integrator creates a delay proportional with ΔV and acts as a memory. Meanwhile, the steady-state error is also minimized. Integrator gain influences the stability as well as system speed entirely. If the gain of this block is large, the control loop oscillates and if this gain is too small the loop will be slow. In Sect. 4.6, the electronic tap-changer system with the modified controller and integral compensator is simulated. Section 4.7 will pay attention to the step-by-step design of compensator in electronic tap-changer with totally novel glance.

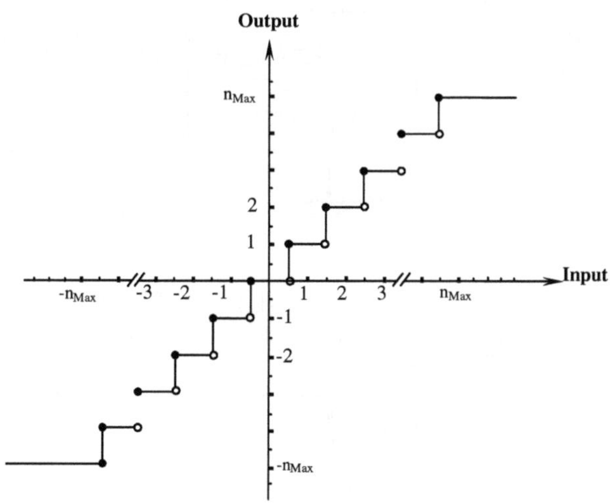

Fig. 4.7 Input–output characteristic of quantizer block

4.6 Simulation of Electronic Tap-Changer with Modified Controller

An electronic tap-changer with its modified controller has been simulated by Simulink software. Figure 4.8 presents the model used in this simulation. To model the power part of the electronic tap-changer, a set of power system blocks of Simulink software has been employed. For modeling the transformer including its secondary taps, three individual transformers with parallel primary windings have been used. Bus Bar block has been utilized for the transformer taps connections. Bi-directional solid-state switches have been modeled by the ideal switch block. It is noted that in this simulation the overall performance of the tap-changer is proposed, and the actual characteristics of the solid-state switches do not change this performance, so the ideal switch has been used in the simulation. Simulation results for three values of the integrator gains have been presented in Figs. 4.9, 4.10 and 4.11. Conditions and parameters used in this simulation are as follows:

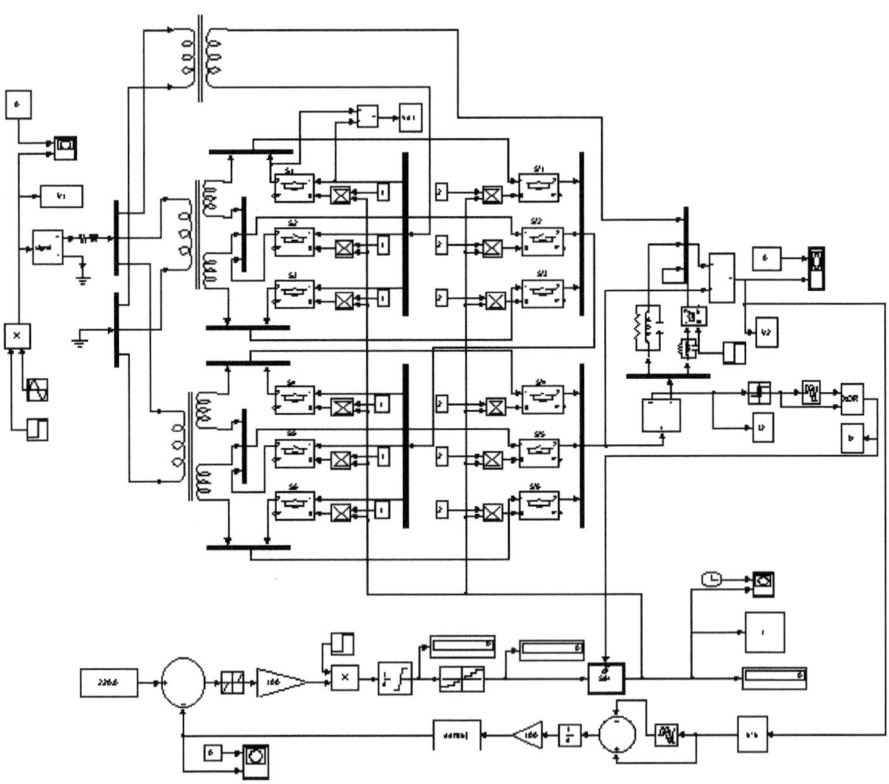

Fig. 4.8 Simulation model of electronic tap-changer with its modified controller

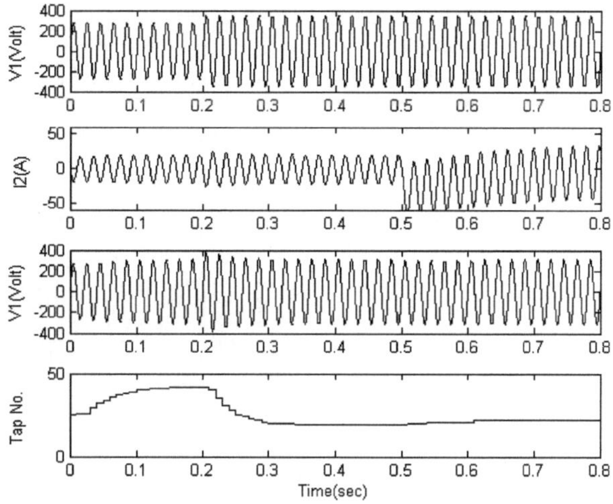

Fig. 4.9 Simulation results of Fig. 4.8 model for integrator gain equal to 10

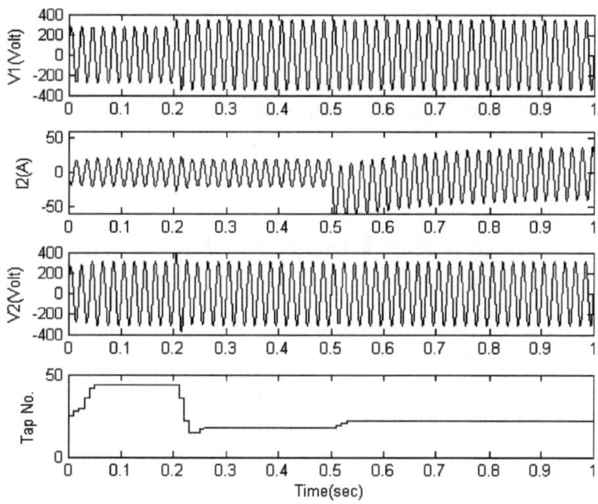

Fig. 4.10 Simulation results of Fig. 4.8 model for integrator gain equal to 30

(a) The input (primary) voltage is applied to the transformer through a series impedance *RL* with $R = 0.1\ \Omega$ and $L = 100\ \mu\text{H}$.
(b) Input voltage is sinusoidal with frequency of 50 Hz, zero phase and rms value of 200 V. The rms value of the voltage suddenly rises to 250 V at $t = 0.2$ s.
(c) The ratio turns of the transformer are as follows:

$$\frac{N_{S_0}}{N_P} = \frac{210}{220}, \quad \frac{N_{S_1}}{N_P} = \frac{2}{220}, \quad \frac{N_{S_2}}{N_P} = \frac{4}{220}, \quad \frac{N_{S_3}}{N_P} = \frac{14}{220}, \quad \frac{N_{S_4}}{N_P} = \frac{28}{220} \tag{4.14}$$

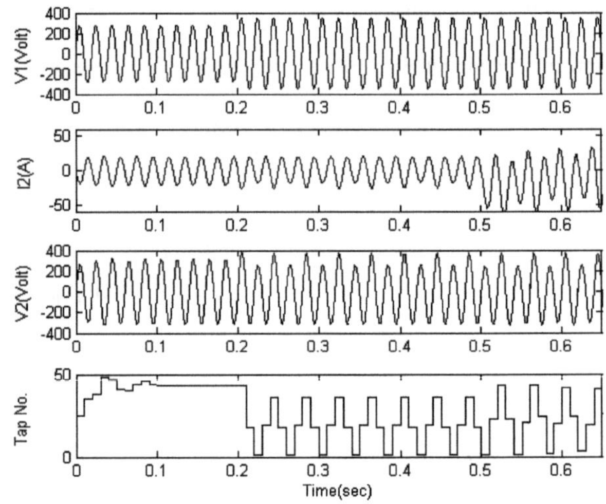

Fig. 4.11 Simulation results of Fig. 4.8 model for integrator gain equal to 100

(d) The load impedance *RLC* is: $R = 15\ \Omega$, $L = 1$ H and $C = 1$ nF (in parallel). As expected, the system becomes faster by rising the integrator gain, but it tends to the instability boundary. Section 4.7 describes how the value of the integrator gain is selected.

4.7 Controller Design for Electronic Tap-Changer

A totally novel procedure is used to design an appropriate controller logically in order to regulate the output voltage of an electronic tap-changer. The design method will be as such that first the problem is simplified by simplifying assumptions, then this simplified problem is solved. The simplifying assumptions are gradually deleted and the necessary modifications will be applied. The simplifying assumptions are as follows:

First Assumption: Waveform of all quantities including primary voltage ($v_1(t)$), secondary voltage (v_2(t)), regulation bus voltage ($v_{2m}(t)$) and load current ($i_2(t)$) are fully sinusoidal with frequency of 50 Hz at all time. Meanwhile, the transient time to approach this sinusoidal state is assumed equal to zero. In other words, the system is always sinusoidal at steady-state.

Second Assumption: The series impedance of transformer (Z_T) and also transmission line impedance up to the regulation bus (SC) and output load impedance are assumed totally ohmic. By this assumption, all quantities including $v_1(t)$, $v_2(t)$, $v_{2m}(t)$ and $i_2(t)$ are in phase.

Third Assumption: It is allowed to change the tap at any time, and the required time for tap changing is zero.

Fourth Assumption: Adjustment and change of the transformer turns ratio by tap changing is totally continuous (over the proposed range). In the other words Δa is assumed equal to 0.

Fifth Assumption: There is a detection method that calculates the values and phases on the sinusoidal quantities with no delay.

Considering assumptions 1, 2 and 4, the governing equations to the system is expressed as follows:

$$V_{2m} = \frac{V_1}{a} - (Z_T(a) + Z_C)I_2, \quad a_{\min} \leq a \leq a_{\max} \tag{4.15}$$

where all quantities in Eq. 4.15, are scalar with zero phases. Therefore, by the above-mentioned assumptions, the system model can be shown as Fig. 4.12a. In this figure, V_1 and I_2 are disturbance inputs, a is input control and finally V_{2m} is the regulated output. By such glace, the system model has been shown in Fig. 4.12b. Considering the above-mentioned simplifying assumptions and Fig. 4.15, the proposed system is a zero degree system with no dynamic.

The aim of designing controller is that V_{2m} approaches the reference value of V_{ref} in the possible shortest time. If it is assumed that quantities V_1 and I_2 are measureable and $Z_T + Z_C$ is known in Eq. 4.15, then a feed-forward block can bring output V_{2m} to V_{ref}. The governing equation to this block is as follows:

$$a = \frac{V_1}{V_{ref} + I_2(Z_T + Z_C)} \tag{4.16}$$

Figure 4.13 presents a tap-changer system with the above-mentioned controller. However, assumption of constant $Z_T + Z_C$ is not correct, because Z_T depends on a. In such a case, feed-forward block of Eq. 4.16 cannot be used, and in addition to the feed-forward there is certainly a need to measure output and also use the feedback path. In Fig. 4.13 the feed-forward path suggests (V_1/V_{ref}) value as a bias

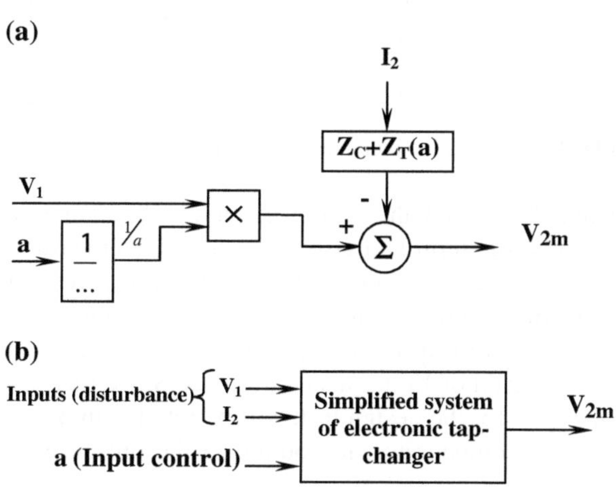

Fig. 4.12 **a** Simplified model of electronic tap-changer, **b** the same as model **a** emphasizing on disturbance inputs, control and also regulated output

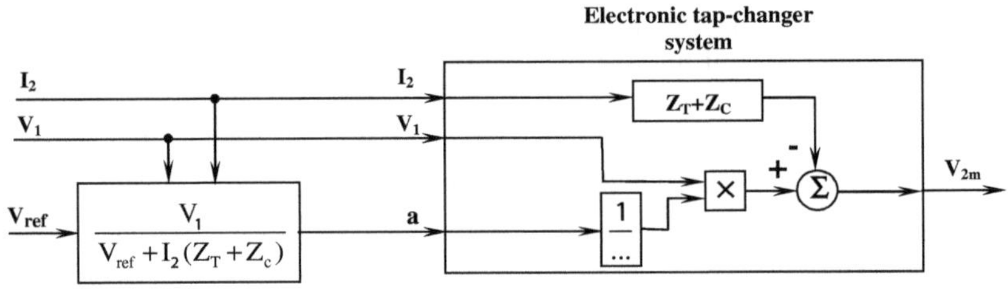

Fig. 4.13 Electronic tap-changer with feed-forward controller

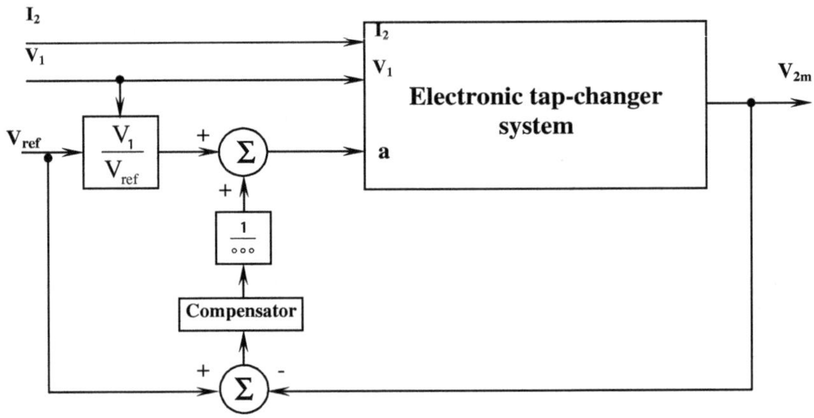

Fig. 4.14 Electronic tap-changer with feed-forward and feedback controller

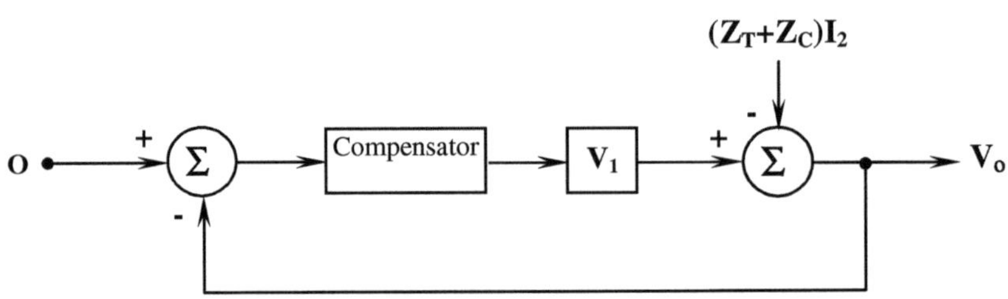

Fig. 4.15 Simplified block diagram of Fig. 4.14

value for a and the feedback path will compensate the error arising from $-(Z_T + Z_C)I_2$ term and also error due to V_1 measurement. To design the compensator in system of Fig. 4.14, this system can be simplified as Fig. 4.15.

Compensator in Fig. 4.15 must be designed as such that the output in the presence of disturbance $(Z_T + Z_C)I_2$ tends to zero quickly. Meanwhile, it is noted that the gain of V_1 has also variations. Since there is no dynamic in block diagram of Fig. 4.15, the design task will be totally simple. It is clear that if a large gain is used as compensator, the output in the presence of the disturbances can tend to zero. A larger gain will make the output closer to zero. If value of this gain is represented by A, the output of block diagram of Fig. 4.15 will be as follows:

$$V_0 = -\frac{I_2(Z_T + Z_C)}{1 + AV_1} \tag{4.17}$$

It is clear from Eq. 4.17 that if value of V_0 is always a positive number smaller than ε, we have:

$$\frac{\frac{|I_2(Z_T+Z_C)|_{max}}{\varepsilon} + 1}{V_{1_{min}}} < A \tag{4.18}$$

If it is necessary that the steady-state error totally tends to zero, an integrator can be used as compensator. In such a case larger integrator gain leads to a larger bandwidth of closed-loop system and also the rate of output (V_0) tendency to zero will be higher. Figure 4.16 presents the simulation results of model, Fig. 4.14, with compensator P (gain). In this simulation it is assumed that at t = 10 s the input voltage V_0 suddenly changes from 200 to 250 V. The load current is considered as a variable with sinusoidal variations of $I_2 = 25 + 1_0 \sin 0.4\pi t$. Simulation has been carried out for two different values of gain A: A = 0.01 and A = 1. A larger gain leads to lower variations of output voltage V_2 against input voltage and also load current and this is expected from Eqs. 4.17 to 4.18. It is noted that variables V_1, V_2 and I_2 are in fact amplitude (rms value) of the actual

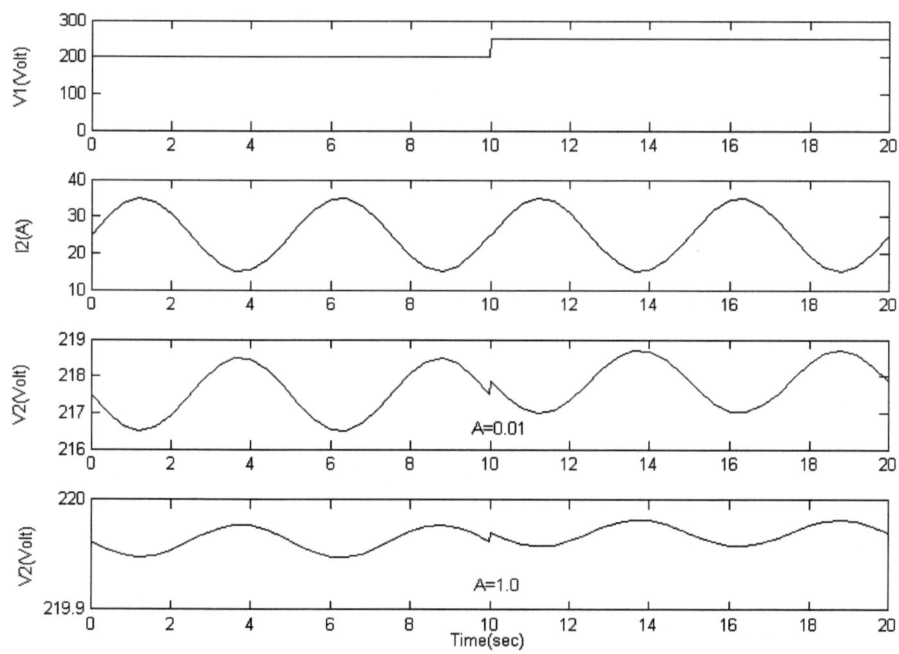

Fig. 4.16 Simulation results of Fig. 4.14 model with compensator *P*

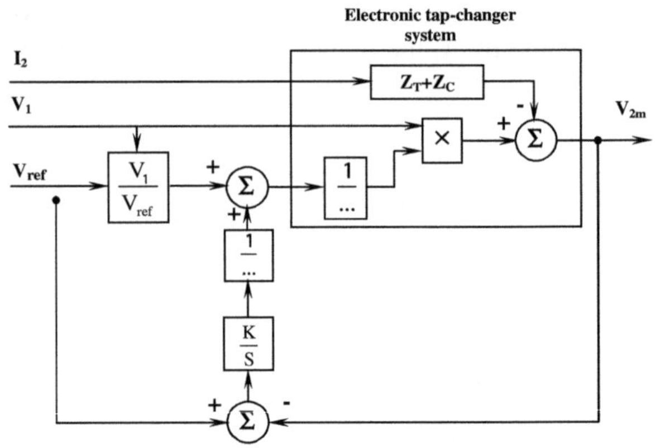

Fig. 4.17 Electronic tap-changer with feed-forward controller and feedback with integral compensator

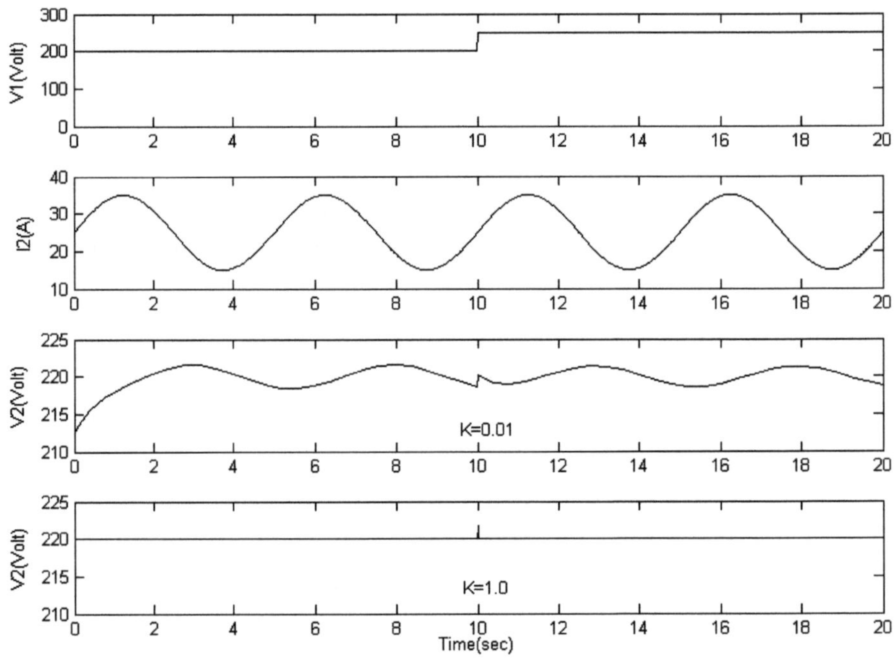

Fig. 4.18 Simulation results of Fig. 4.17 model with integral compensator

sinusoidal variables. Meanwhile, in this simulation $V_{ref} = 220$ V, and by using compensator P the steady-state error becomes non-zero, but by increasing the gain value this error diminishes.

Figure 4.17 exhibits an electronic tap-changer with feed-forward controller and also feedback with internal compensator. In this case, the closed-loop system time constant is $1/KV_1$, where K is the integrator gain. Figure 4.17 has been simulated for two integrator gains 0.01 and 1. Figure 4.18 shows the results of the simulation. The disturbance input signals in this simulation corresponds with the signals in Fig. 4.16. It is seen in Fig. 4.18 that by increasing the integrator gain, the speed of the system rises. Meanwhile, the steady-state error is zero in both cases, because of integrator application.

So, design of the controller using the simplifying assumptions is ended, Now each simplifying assumptions which is not set permanently is removed and then necessary modifications is applied to the controller of Fig. 4.17.

First the assumption of in phase quantities $v_1(t)$, $v_2(t)$, $i_2(t)$ and $v_{2m}(t)$ are removed. In the order words, Eq. 4.2 is concealed. In such a case, series impedances Z_T and Z_C and also load impedance are not necessarily ohmic and Eqs. 4.1 and 4.5 are in phasor forms. In this case, responsibility of the controller is that $|V_{2m}|$ tends to V_{ref}. If it is assumed: $\widehat{V}_1 = V_1 e^{j0}$, $\widehat{I}_2 = I_2 e^{j\theta_I}$, $\widehat{Z}_T + \widehat{Z}_C = Ze^{j\theta_z}$ then it can be easily shown that the appropriate value of a for $|V_{erf}| = V_{2m}$ is as follows:

$$a = \left(\frac{I_2 Z\cos(\theta_z + \theta_I)}{V_1} + \frac{\sqrt{V_{ref}^2 - Z^2 I_2^2 + Z^2 I_2^2 \cos^2(\theta_z + \theta_I)}}{V_1} \right)^{-1} \tag{4.19}$$

It is clear if θ_1 and θ_2 be zero in Eq. 4.19, this equation will convert to Eq. 4.16. Suppose in this case parameters Z and θ_Z are known and quantities V_1, I_2 and θ_1 are measurable and available, $|V_{2in}|$ can be regulated and controlled by a feed-forward block as open-loop. Such system has been presented in Fig. 4.19.

Here, it is also assumed that Z and θ_Z are not the actual values and thus measuring output and feedback path are obligatory. Z and θ_Z in this figure are unknown and can be considered equal to zero or totally the feed-forward part is neglected.

In this case, similar to the previous case, if compensator be a large gain, $|V_{2m}|$ tends to V_{ref} in the presence of $I_2(Z_T + Z_C)$. To prove this, by using Eqs. 4.1–4.5, the relationship between $|V_{2m}|$ and other parameters of the system can be expressed as follows:

$$|V_{2m}| = \sqrt{\left(\frac{V_1}{a} - Z I_2 \cos(\theta_z + \theta_I)\right)^2 + (Z I_2 \sin(\theta_z + \theta_I))^2} \tag{4.20}$$

Now if compensator has a large gain A, then without taking into account the feed-forward path, a can be obtained as follows:

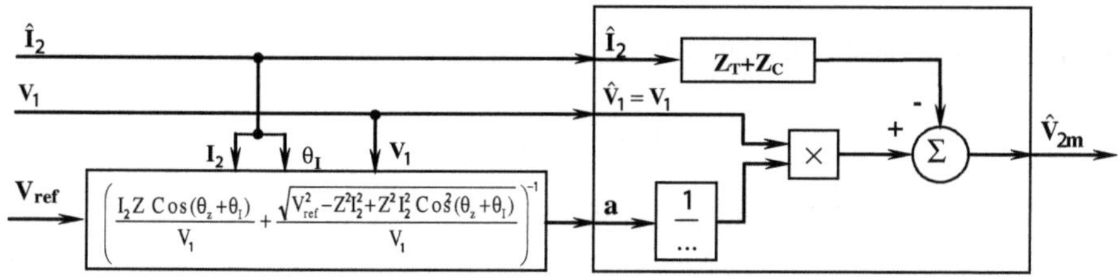

Fig. 4.19 Electronic tap-changer system with feed-forward controller in the case of non-ohmic impedances

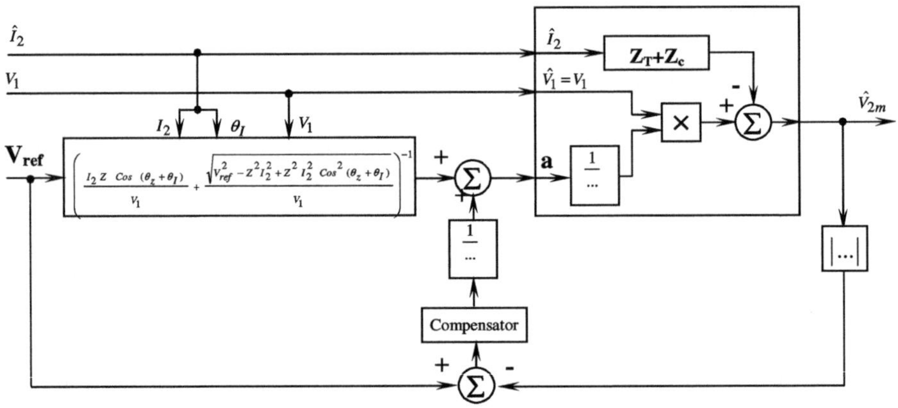

Fig. 4.20 Electronic tap-changer system with feed-forward and feedback controller in the case of non-ohmic impedances

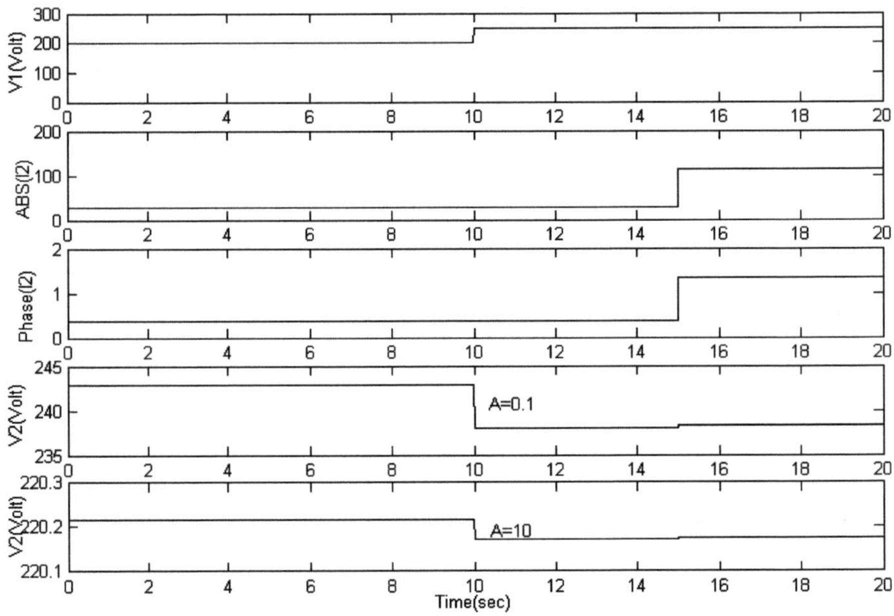

Fig. 4.21 System simulation results of Fig. 4.20 model for compensator P

$$A(V_{\text{ref}} - |V_{2m}|) = \frac{1}{a} \tag{4.21}$$

Combination of (4.20) and (4.21) yields:

$$|V_{2m}| = \sqrt{(A(V_{\text{ref}} - |V_{2m}|)V_1 - Z I_2 \cos(\theta_z + \theta_I))^2 + (Z I_2 \sin(\theta_z + \theta_I))^2} \tag{4.22}$$

Since A is large, other terms of the right hand side of Eq. 4.22 can be neglected and following approximated equation is obtained:

$$|V_{2m}| \cong |A(V_{\text{ref}} - |V_{2m}|)V_1| \tag{4.23}$$

Now Eq. 4.23 is rewritten as follows:

$$\frac{1}{AV_1} = \frac{|V_{\text{ref}} - |V_{2m}||}{|V_{2m}|} \tag{4.24}$$

Equation 4.24 shows that considering the large gain A, value of $|V_{\text{ref}} - |V_{2m}||$ must be small enough to satisfy Equality (4.24). A larger A makes $|V_{\text{ref}} - |V_{2m}||$ closer to zero and this is the proposed aim. Of course, considering Eq. 4.22, it is noted that the closed-loop system will be stable if the following is satisfied:

$$\left|\frac{V_1}{a}\right| = |A(V_{\text{ref}} - |V_{2m}|)V_1| > |ZI_2\cos(\theta_z + \theta_I)| \tag{4.25}$$

Since the drop of the series impedances is usually very smaller than the induced voltage in the secondary, Eq. 4.15 is satisfied in the actual conditions; thus, the system is stable.

An integrator can be used in place of a large gain A as compensator. In this case integrator gain will determine the system speed. Figures 4.21 and 4.22 show the system simulation results due to the system of Fig. 4.20 for compensator P and

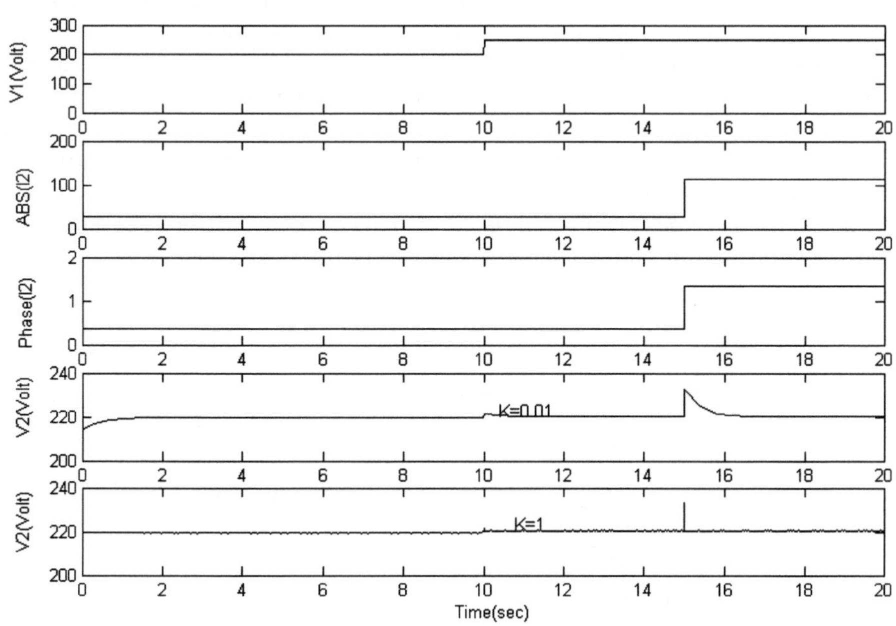

Fig. 4.22 System simulation results of Fig. 4.20 model for integral compensator

compensator I respectively. In this simulation the input voltage (V_1) is changed from 200 to 250 V at $t = 10$ s and load current phasor jumps from 25 + j10 to 25 + j110 at $t = 15$ s. Meanwhile, sum of series impedances 0.3 + j0.1 is considered in this simulation.

Now the assumption of continuous a ($\Delta a = 0$) is concealed. In other words, assumption (4) is neglected. In this case, a limited individual values will be permissible, and the model of the system is as shown in Fig. 4.23.

Certainly, quantizing a creates an un-compensable error in the output; however, this error (which is an inherent error of the system) must not mistake the controller system. In other words, this error must not excite the controller, otherwise; system oscillates. This is the reason that a Dead-Zone block is necessary after comparison of $|V_{2m}|$ and V_{ref}. Therefore, the controller system must be modified as shown in Fig. 4.24.

Value of DB is defined referring to the peak inherent error of the electronic tap-changer (see Sect. 4.5). If this DB is smaller than the peak of the possible error, system will oscillate; if it is larger, the peak of the possible error leads to errors larger than the inherent error of the system. Figure 4.25 presents the simulation results of Fig. 4.24. In this simulation, the similar input disturbance signals conditions (input voltage V_1 and load current I_2) are simulated in Figs. 4.21 and 4.22. The output voltage in this figure has been plotted for two values of DB: 1.5 and 10. It is seen that in this case, in spite of existing integral compensator, the output steady-state error is not zero, and this error is increased by increasing DB.

Now, assumption of sinusoidal quantities of the system (first assumption) is concealed. This is due to two factors, first voltage $v_1(t)$ or current $i_2(t)$ is non-sinusoidal and second transient times to reach the steady-state sinusoidal mode; the later is created because of the load variations, or $v_1(t)$ variations and/or variations of a. If in the method for amplitude and phase detection of the sinusoidal quantities can separate the fundamental sinusoidal component from other signals and harmonics, non-sinusoidal quantities will not produce any problem. The reason is that Eqs. 4.1 and 4.5 are true for fundamental component, and controller has been designed based on the design equations. Methods of amplitude and phase of the sinusoidal quantities will be discussed in Sect. 4.8.

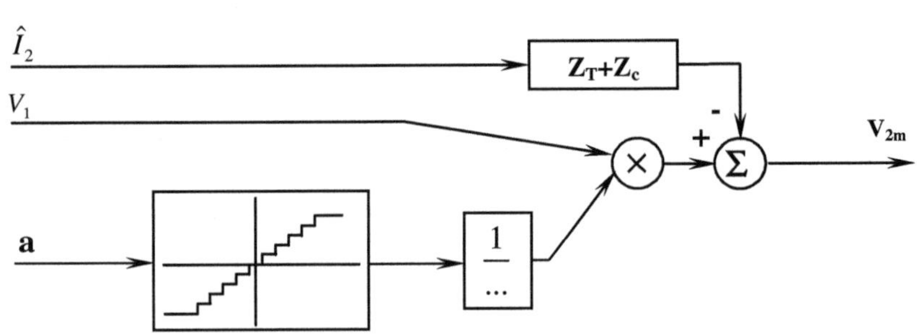

Fig. 4.23 Electronic tap-changer model considering quantized a

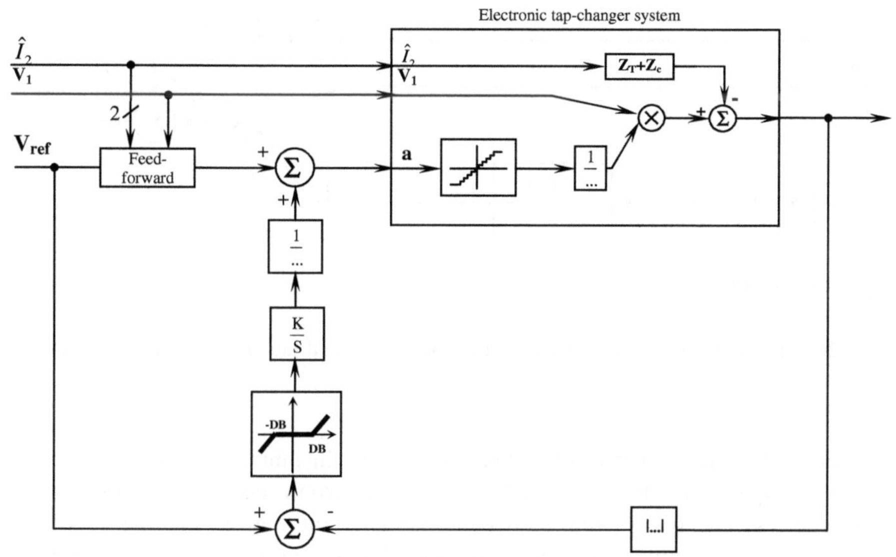

Fig. 4.24 Electronic tap-changer with modified controller and added DB block

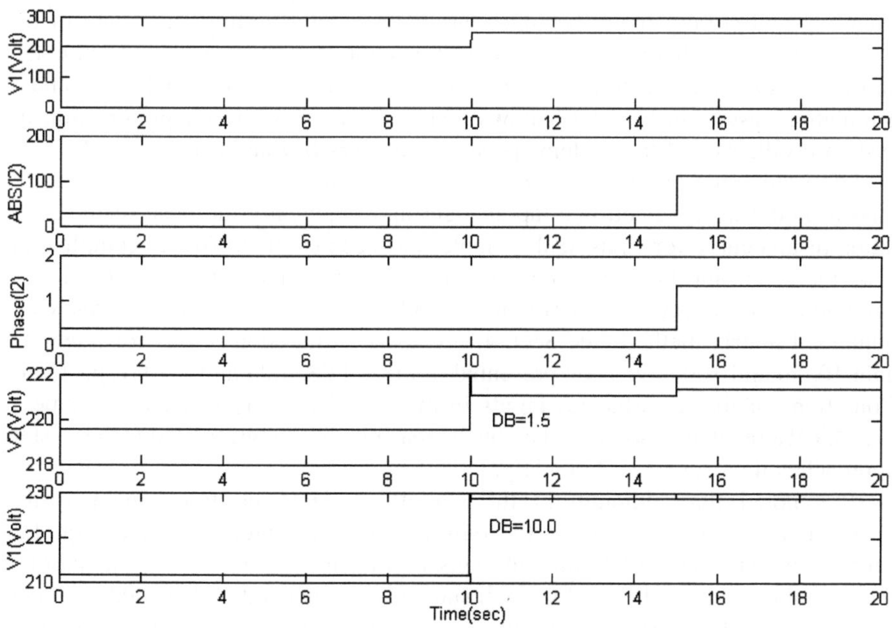

Fig. 4.25 Simulation results of Fig. 4.24 model for DB equal to 1.5 and 10

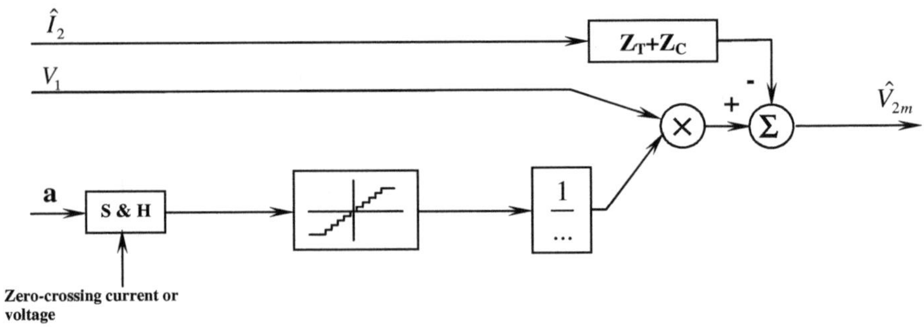

Fig. 4.26 Electronic tap-changer model with quantizing *a* and tap changing at instants of zero crossing load current

Finally, the 3rd assumption (tap changing at any instant) is canceled. To reduce switching losses and stresses in the designed electronic tap-changer, only at the zero load current (or voltage) crossing instants the tap is changed. By this assumption, the system model can be presented as shown in Fig. 4.26. By limiting the tap changing time to the instants of zero-crossing load current, a delay is practically added to the system and controller may oscillate. The reason for such oscillation is the delay that imposed by the sampling and storing block to the system.

In this case, in order to have non-oscillating closed-loop system, the integrator gain must not be selected too large, so the rate of signal *a* changes before the sampling and storing block becomes larger than the sampling frequency of this block; otherwise, there is no guarantee for stability of the system. Figure 4.27 shows the simulation results of Fig. 4.26 for integrator gains 0.001, 0.01 and 1. Simulation results in Fig. 4.27 shows that by increasing the integrator gain, the system oscillates and the system speed cannot rises beyond a define limit.

All simulations presented in this section have been obtained considering the phasor model of the electronic tap-changer and practically the push of the simulation results curve for sinusoidal variables has been studied. Now to complete the task, the electronic tap-changer is simulated using a time variant model. In this simulation, to detect the amplitude and phase of the sinusoidal variables, a numerical matrix method (see Sect. 4.8) is used. Simulation results for integrator gain 10, 30 and 100 have been presented in Figs. 4.28, 4.29 and 4.30 respectively. Conditions of the disturbances inputs in this simulation are exactly the same as Fig. 4.9. Here, it is also seen that by increasing the integrator gain, the system approaches the oscillating boundary.

It is noted that although in the simplified electronic tap-changer system (Fig. 4.15) it is possible to use PI controller, the presented system in Fig. 4.26 cannot utilize this controller after adding sampling and storing block. The reason is that signal *a* as input of this block will suddenly be changed by the sudden change of the input voltage or load current. So, condition of slow changes of the input signal of sampling and storing block cannot be realized at all.

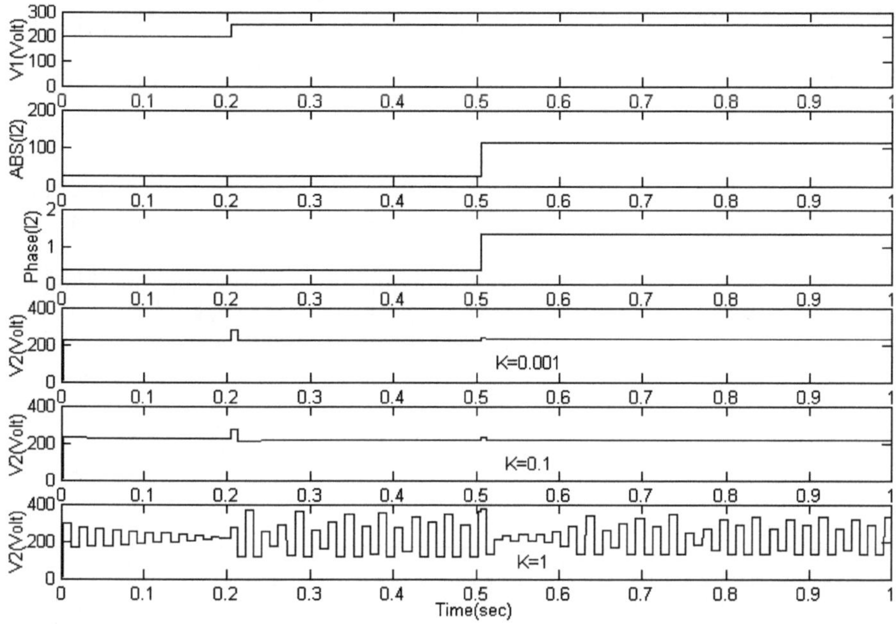

Fig. 4.27 Simulation results of Fig. 4.26 model for three integrator gain: 0.001, 0.1 and 1

Fig. 4.28 Simulation results of electronic tap-changer for integrator gain equal to 10

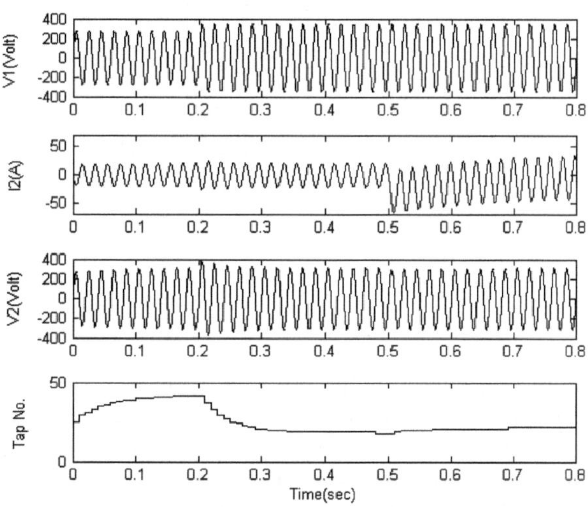

4.8 Amplitude and Phase Measurement of a Sinusoidal Variable Algorithm

Those discussed in Sect. 4.7 on the electronic tap-changer control system were based on the availability of a very fast algorithm for extracting the amplitude and phase of the sinusoidal variables data. For instance, in the control block diagram of

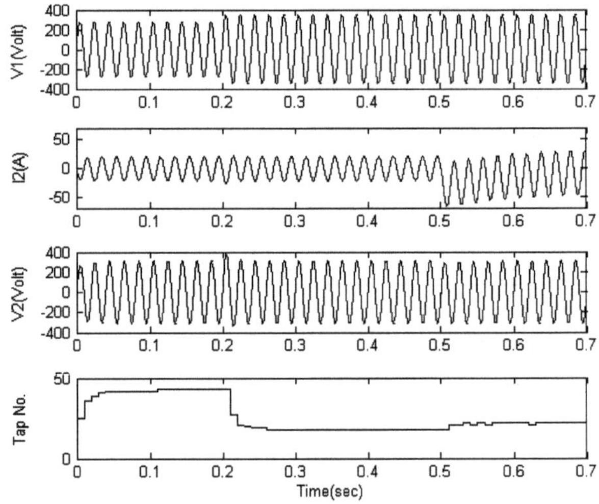

Fig. 4.29 Simulation results of electronic tap-changer for integrator gain equal to 30

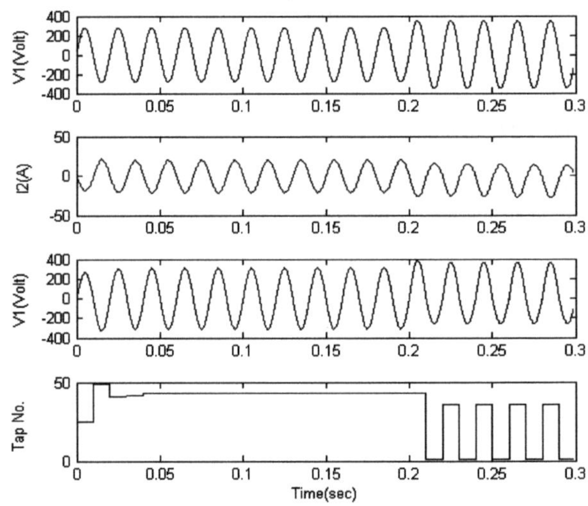

Fig. 4.30 Simulation results of electronic tap-changer for integrator gain equal to 100

Fig. 4.24, data of amplitude and phase of variable I_2, amplitude of variable V_1 for feed-forward and amplitude of variable V_{2m} for feedback section are used. Extracting this data needs an appropriate algorithm. In the following sub-section, some methods are reviewed [5–8] for this task and then these methods are compared. Finally, a suitable method for the electronic tap-changer system is selected.

4.8.1 Use of Sinusoidal Variable Peak

The simplest method for extracting the amplitude of a sinusoidal variable is the measurement of its peak value. There are different methods to find the instant of the peak of the sinusoidal variable. One of the commonest methods is calculation of the derivative of the sinusoidal variable and finding the time at which this derivative becomes zero.

It is noted that measuring the amplitude of the sinusoidal variable using this method will have maximum up to an half cycle delay. Meanwhile, this method is seriously sensitive to the noise or the existing parasitic harmonics in the measured sinusoidal variable, therefore, except in very simple applications, it is not used. In addition, the discussed method has no phase measurement capability.

4.8.2 Evaluation of rms Value of Sinusoidal Variable

The next method for extracting the amplitude of sinusoidal variable is the evaluation of its rms value. This method involves integral calculation and basically is an averaging method. Measuring the rms value of a sinusoidal variable can be done in every cycle, or half cycle or quarter of cycle. Certainly, a shorter integration time will increase the measurement rate; however, its sensitivity to the noise and distortions will be more.

Another method is evaluation of the instantaneous rms value. In this method, rms value of the sinusoidal variable in a time window is calculated. Limits of the window is between t (measuring instant) and t − 0.5 T or $t - T$ (T is the period of sinusoidal variable). Although instantaneous rms value leads to a shorter delay compared to the rms value evaluation in every cycle or half a cycle, in the worse case of the half cycle conditions, there will be delay in the precise evaluation of the rms value.

It is noted that to calculate the rms value, the data of zero crossing of the sinusoidal variable is required. Noise may influence these times and leads to error in measuring rms value. In addition, this method has no capability of measuring the phase.

4.8.3 90° Delay of Sinusoidal Variable

Block diagram of Fig. 4.31 shows that how this method is applied. The method is based on the following trigonometry equality:

$$A^2\sin^2(\omega t+\theta)+A^2\sin^2\left(\omega t+\theta-\frac{\pi}{2}\right)=A^2 \tag{4.26}$$

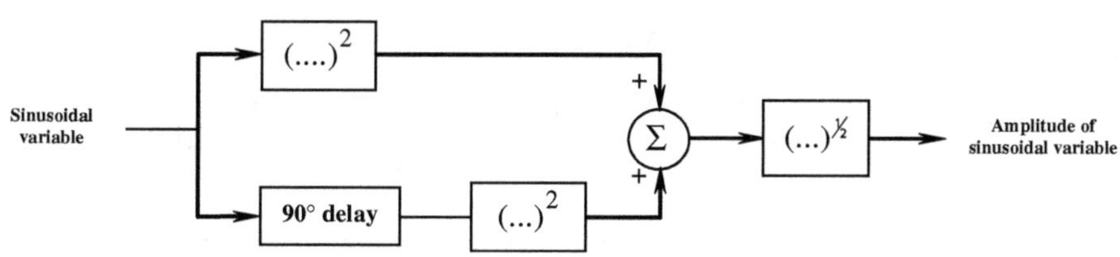

Fig. 4.31 Block diagram of detection of amplitude of sinusoidal variable by its 90° delay

This method also has 90° delay due to the use of 90° delay block. In addition, the discussed method has no capability of measuring the phase.

4.8.4 Fourier Transform of Sinusoidal Variable

Fourier series transform of a sinusoidal variable is a method that extracts amplitude and phase of the variable using fast algorithms. Of course, this method also is typically an averaging method: thus, it has delay. The very important advantage of this method is that this method is not sensitive to additional harmonics because every harmonic is basically evaluated separately, while existing additional harmonics in the sinusoidal variable created error in the previous methods.

4.8.5 Using Phase Locked Loop (PLL)

Figure 4.32 shows the block diagram of this method. Block of generating signals α and β creates signals V_α and V_β from input variable (u). The governing equations for this block is as follows:

$$\begin{pmatrix} V_\alpha \\ V_\beta \end{pmatrix} = u\begin{pmatrix} -e^{-j\pi/2} \\ 1 \end{pmatrix} \tag{4.27}$$

Then dq transform block tries to place d-axis on vector $V_\alpha + jV_\beta$. θ in PLL varies until V'_{q} approaches zero, therefore, θ represents the phase of the sinusoidal variable and V'_{d} contains data relevant to the amplitude. Governing equations on transform block to dq frame are as follows:

$$\begin{pmatrix} V'_{\mathrm{d}} \\ V'_{\mathrm{q}} \end{pmatrix} = \begin{pmatrix} \cos\theta & \sin\theta \\ -\sin\theta & \cos\theta \end{pmatrix}\begin{pmatrix} V_\alpha \\ V_\beta \end{pmatrix} \tag{4.28}$$

By changing the phase of the input sinusoidal variable, PLL exits from locked state temporarily and after passing the transient time, it will again go to the locked

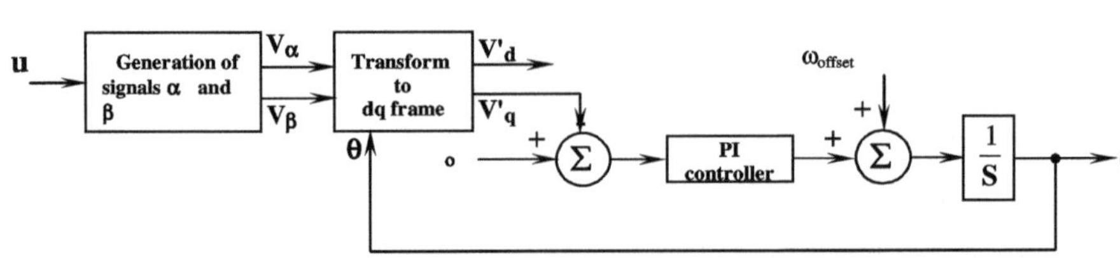

Fig. 4.32 Block diagram of detecting amplitude and phase of sinusoidal variable using PLL

state. This method has practically this transient as a delay for estimation of the amplitude and phase.

4.8.6 Using Numerical Matrix Method

Contrary to the previous methods, this method has a shorter delay and contains amplitude and phase data. In this method, it is assumed that the important harmonic components of the measured sinusoidal variable are defined. The basis of this method is that the main signal is sampled by a sampler block and amplitude and phase data are extracted by some calculations on a define number of the samples. For example, if it is assumed that the sinusoidal variable in addition to the fundamental component contains the 5th harmonic; then, the governing equations on the samples of signal can be written as follows:

$$\begin{aligned} u_{\text{present}} &= |u_1|\cos(\omega_1 t + \theta_1) + |u_5|\cos(\omega_5 t + \theta_5) \\ u_{\text{past}_1} &= |u_1|\cos(\omega_1 t + \theta_1 - \omega_1 T_s) + |u_5|\cos(\omega_5 t - T_s\omega_5 + \theta_5) \\ u_{\text{past}_2} &= |u_1|\cos(\omega_1 t - 2\omega_1 T_s + \theta_1) + |u_5|\cos(\omega_5 t - 2\omega_5 T_s + \theta_5) \\ u_{\text{past}_3} &= |u_1|\cos(\omega_1 t - 3\omega_1 T_s + \theta_1) + |u_5|\cos(\omega_5 t - 3\omega_5 T_s + \theta_5) \end{aligned} \tag{4.29}$$

where $|u_1|$ is the amplitude of the fundamental component (50 Hz); $|u_5|$ the amplitude of the 5th harmonic (250 Hz); θ_1 the fundamental harmonic phase; θ_5 the 5th harmonic phase; ω_1 the fundamental harmonic angular frequency; ω_5 the 5th harmonic angular frequency; t time; T_s the sampling period; u_{present} the sample signal at t (present sample); and $u_{\text{past1-3}}$ is the sample signal at $t - T_s$, $t - 2T_s$, $t - 3T_s$ (three previous samples).

There are four equations in (4.29). It is clear that if the number of harmonics considers for signal u is larger or lower than 4, the number of the required samples will vary and finally the number of the equations is lower or higher in the same order. In fact, every harmonic adds two unknown values (amplitude and phase) to the unknown values, so to find the unknown values, it is necessary to have two equations for every harmonic and thus two samples.

Now to find the unknown values in Eq. 4.29, those equations are arranged as the following matrix form:

$$\overbrace{\begin{pmatrix} V_{\text{present}} \\ V_{\text{past}_1} \\ V_{\text{past}_2} \\ V_{\text{past}_3} \end{pmatrix}}^{B} = \overbrace{\begin{pmatrix} 1 & 0 & 1 & 0 \\ \cos\omega_1 T_s & \sin\omega_1 T_s & \cos\omega_5 T_s & \sin\omega_5 T_s \\ \cos 2\omega_1 T_s & \sin 2\omega_1 T_s & \cos 2\omega_5 T_s & \sin 2\omega_5 T_s \\ \cos 3\omega_1 T_s & \sin 3\omega_1 T_s & \cos 3\omega_5 T_s & \sin 3\omega_5 T_s \end{pmatrix}}^{A} \overbrace{\begin{pmatrix} |u_1|\cos\theta_1 \\ |u_1|\sin\theta_1 \\ |u_5|\cos\theta_5 \\ |u_5|\sin\theta_5 \end{pmatrix}}^{X} \tag{4.30}$$

Now the unknown values matrix (x) is easily obtained as follows:

$$\begin{pmatrix} x_1 \\ x_2 \\ x_3 \\ x_4 \end{pmatrix} = \overbrace{\begin{pmatrix} |u_1|\cos\theta_1 \\ |u_1|\sin\theta_1 \\ |u_5|\cos\theta_5 \\ |u_5|\sin\theta_5 \end{pmatrix}}^{X}$$

$$= \overbrace{\begin{pmatrix} 1 & 0 & 1 & 0 \\ \cos\omega_1 T_s & \sin\omega_1 T_s & \cos\omega_5 T_s & \sin\omega_5 T_s \\ \cos 2\omega_1 T_s & \sin 2\omega_1 T_s & \cos 2\omega_5 T_s & \sin 2\omega_5 T_s \\ \cos 3\omega_1 T_s & \sin 3\omega_1 T_s & \cos 3\omega_5 T_s & \sin 3\omega_5 T_s \end{pmatrix}^{-1}}^{A^{-1}} \overbrace{\begin{pmatrix} u_{\text{present}} \\ u_{\text{past}_1} \\ u_{\text{past}_2} \\ u_{\text{past}_3} \end{pmatrix}}^{B} \tag{4.31}$$

Since sampling time T_s and angular frequencies ω_1, ω_2 are known, matrix A^{-1} and so unknown values of x_1, x_2, x_3 and x_4 are calculable. Now, amplitude and phase of every harmonic can easily be evaluated by the known values of the matrix coefficients as follows:

$$\begin{aligned} \theta_1 &= \text{Arc tan}\left(\frac{x_2}{x_1}\right) \\ |u_1| &= \frac{x_1}{cos\,\theta_1} \\ \theta_5 &= \text{Arc tan}\left(\frac{x_4}{x_3}\right) \\ |u_1| &= \frac{x_3}{cos\,\theta_5} \end{aligned} \tag{4.32}$$

In this method the delay time for measuring the data of the amplitude and phase is obtained as follows:

$$\text{Time delay of measurement} = \text{twice of harmonics number} \times T_s \tag{4.33}$$

Therefore, by shortening the period of the sampling time (T_S), the measuring delay time can be shortened. Of course it must be noted that by shortening the sampling time period, the error appears more likely.

4.8.7 Comparison Between Proposed Methods and Selection of an Appropriate Method

Table 4.1 compares the methods discussed in the previous sub-sections. Criteria for comparison in this table are:

(1) the longest delay time in measurement, certainly a more appropriate method has a shorter delay. In other words, it can extract the amplitude and phase of the sinusoidal variable quickly;

Table 4.1 Comparison of different methods in measuring amplitude and phase of a sinusoidal variable

No.	Evaluation criterion Method name	Longest measurement delay	Capability of calculation		Sensitivity to additional harmonics	Complexity of method	Notes
			Amplitude	Phase			
1	Using peak of sinusoidal variable	An half cycle	Yes	No	Yes	Very simple	–
2	Evaluation of rms value of sinusoidal variable	An half cycle	Yes	No	Yes	Simple	Rms value calculated over an half cycle
3	90° delay of sinusoidal variable	Quarter of cycle	Yes	No	Yes	Simple	–
4	Fourier transform of sinusoidal variable	One cycle	Yes	Yes	No	Complicated	Sampling over one period
5	Using PLL	t_s	No	Yes	Yes	Complicated	t_s: necessary time for PLL
6	Numerical matrix	$2NT_s$	Yes	Yes	No	Average	N: no of harmonics T_s: sampling period

(2) capability of the extracting relevant data to the amplitude and phase which are necessary in the proposed application;
(3) the sensitivity to the additional harmonics which is important in the choice of the appropriate method due to the existing harmonics in the voltage and current of public electricity supply; and finally
(4) the last criterion is simplicity of the implementation of the measurement method.

Referring to Table 4.1, it is seen that the evaluation of the numerical matrix method is more appropriate for application in the electronic tap-changer control system.

References

1. Kasztenny B, Rosolowski E, Izykowski J, Saha MM, Hillstrom B (1998) Fuzzy logic controller for on-load transformer tap-changer. IEEE Trans Power Deliv 13(1):164–170
2. Faiz J, Siahkolah B (2006) Differences between conventional and electronic tap-changers and modifications of controller. IEEE Trans Power Deliv 21(3):1342–1349
3. Bassett DL (1993) Control of tap changer under load transformers through the use of programmable logic controllers. IEEE Trans Power Deliv 8(8):1759–1765
4. Donda V, Hiskens IA (2002) Analysis of limit cycle stability in a tap-changing transformer, IEEE ISCAS 02, May 2002, pp V/693–V/696
5. Chung J, Powers EJ, Grady WM, Bhatt SC (1999) Adaptive power-line disturbance detection scheme using a prediction error filter and a stop-and-Go CACFAR detector. IEEE ICASSP 99 3:1533–1536
6. Chung J, Powers EJ, Grady WM, Bhatt SC (2002) Power disturbance classifier using rule-based method and wavelet packet hidden Markov model. IEEE Trans Power Deliv 17(1):233–241
7. Song HS, Nam K, Mutschler P (2002) Very fast phase angle estimation algorithm for a single phase system having sudden phase angle jumps, IEEE-IAS, Pittsburg
8. Fitzer C, Barnes M, Green P (2004) Voltage sag detection technique for a dynamic voltage restorer. IEEE Trans Ind Appl 40(1):203–212

Chapter 5
Electronic Tap-Changers as Custom Power Tools

5.1 Background

5.1.1 Introduction to Power Quality

Electrical power generator units supply small and large electrical energy consumers in the most economical way, as such that the delivered electrical energy has acceptable reliability and quality [1]. But for many decades, electricity suppliers had concentrated on generation, transmission and distribution of electrical energy to the consumers. This was while reliability and quality were not taken into account seriously [2]. The reason was that those days lighting was the most important load of consumers. Of course electrical motors loads have been gradually grown as major loads of the power systems. However, consumers could wait hours for the re-connection of electricity.

As time passes by, rising the industrialization trend and the increasing dependency of the daily life on the electrical energy, expectation from electrical energy reliability increased. Reliability is defined as total hours in 1 year that electrical energy is available. Demand for high reliability leads to structural changes in the transmission (connection of networks) and distribution (use of several feeders and enhancement of maintenance level and equipment services) of networks. These changes and modifications in the developed countries result in only once or twice long term (longer than 10 s) electricity block-out in a year [3].

In the second half of the 1980s, consumers' demands changed again and now the trend of these changes is increasing. The reason behind the changes in the consumers' demands is the application of new electronic equipment. These new loads include microprocessor systems, computers, advanced communication devices, automatic controllers, robots, automatic control of product processes, drives of ac and dc motors etc [3]. These equipments are very sensitive and even a very short block-out (a few cycles) is unacceptable to the consumers. Therefore, the concept of reliability as total available hours of electrical energy has been abolished. For example, 2,400 block-outs, each 0.25 s duration, in total will be

J. Faiz and B. Siahkolah, *Electronic Tap-changer for Distribution Transformers*, Power Systems, DOI: 10.1007/978-3-642-19911-0_5,

10 min, but each of these block-outs can cause many problems in the above-mentioned electronic equipment and this in turn can cause great damage to the equipment. Therefore, the concept of "power quality" was proposed, which certainly has wider dimensions compared to reliability. In the late 1980s, the term power quality had been very often used in the electrical energy industry and it was also one of the proposed topics in EPRI. Of course [4], from historical point of view, the first use of the term power quality was in the introduction of a paper published in 1968. In this paper, studies had been presented on the required power specifications in electrical and electronic equipment by USA navy force [2].

Nowadays, the fact that the power quality is a very important aspect in delivering electrical power to the consumers is universally agreed upon, particularly after the second half of the 1990s up [2].

5.1.2 Definition of Power Quality

There are different definitions for power quality. Each definition looks at the problem from a particular viewpoint. Basically, the topic of power quality is important for the three following groups:

(a) Electricity companies which are responsible to supply electrical energy to consumers.
(b) Consumers, particularly industrial and commercial subscribers.
(c) Designers and manufacturers of electronic equipment and devices.

From viewpoint of electricity companies, power quality is synonymous with reliability. On the contrary, designers and manufactures of electronic equipment define power quality as appropriate performance of the above-mentioned equipment based on a suitable input. This definition can have different meaning for various devices and equipment and also different manufactures [4].

However, power quality is a problem arising from the needs of consumers and thus their viewpoints in this matter plays a major role. Based on this, power quality definition will be as follows [5]:

Power quality is the evaluation of the supply of the sensitive devices of the subscribers as such that they operate properly. So it is clear that the term power quality is a vast concept and includes many types of electromagnetic phenomena. These phenomena are as follows:

(a) Transient phenomena of impulse and oscillation.
(b) Short term voltage falling and rising.
(c) Waveforms distortions (harmonics, notch, noise,…).
(d) Frequency change.
(e) Flicker.

None of the above-mentioned phenomena are new concepts, but engineers and researchers currently attempt to encounter this concept from a general and systematic viewpoint and not as individual phenomena and miscellaneous, and this can be considered totally a new concept. According to the power quality definition, the power quality problems are defined as follows:

Any deviation from the rated values in the network, which cause mal-operation and improper performance of sensitive electronic devices, is power quality problem. It must be noted that the reliability (total full block out hours) is hidden in the power quality concept, in the other words it can be expressed that [5]:

Power quality = waveform quality plus electricity reliability.

5.1.3 Quality of Voltage and Current

The most important criticism against the "power quality" terminology is that there is no quality for a physical quantity such as "power". Some other terms to describe qualitative relationships between electricity suppliers and consumers are proposed below which will show that the "power quality" term is so far the best and most rational term [3].

- *Voltage quality* Voltage quality expresses the level of voltage deviation from its ideal value (fully sinusoidal voltage with a fixed frequency and amplitude). Considering that basically distribution networks are only capable to control the voltage quality and have no control on the loads current, voltage quality can be considered as a term for expressing the products generated by the electricity suppliers and delivered to the consumers.
  This term which is particularly used in the European countries, only covers the technical aspects of voltage. It must be noted that there is always a close relationship between voltage and current in any power network. Although generators generate a full approximated voltage waveform, the load current passing the network impedance can cause various disturbances in the voltage. For example [4]:

  (a) Short circuit current causes voltage reduction or zero voltage;
  (b) Distorted currents due to the harmonic loads lead to distorted voltage waveform.

However, final attention is paid to the voltage, the existing phenomena in the current (current quality) must also be taken into account in order to understand voltage distortions. So, term "power quality" is more comprehensive.

- *Current quality* This term describes current deviation from the ideal current. In fact current quality explains the impact of the consumer on the network.
- *Power quality* This term proposes the combination of voltage and current qualities.
- *Quality of supply* This term contains the technical parts of voltage quality and service quality. The purpose of the service quality is that how quick is the response to the complaints of the consumers, and clarification of structure and tariffs.
- *Quality of consumption* This term contains the technical parts of the current quality and the non-technical section of the service quality.

- *Electromagnetic compatibility (EMC)* This term is used in International Electrotechnical Commission (IEC) standards and generally includes the interface between the devices and between device and supply. There are two basic concepts in EMC:

(a) *Emission*: is the electromagnetic pollution generated by equipment
(b) *Immunity*: is the ability of a device to confront and stand the electromagnetic pollution.

In the figure of these concepts current quality will be in relation with the emission and voltage quality will deal with the immunity. In total this term is not completely clear and the use of EMC in the power quality is not recommended.

5.1.4 Reasons for Importance of Power Quality

As described in Sect. 5.1.1, the most basic reason of importance of power quality must be searched around the wide application of new electronic equipment and their high sensitivity. The importance of the issue of power quality is introduced precisely and in more details in following part [6, 7]:

(a) Sensitivity of the present electronic equipment to different types of voltage disturbances is more than that of the previous equipment. For industrial consumers this is explained as follows:

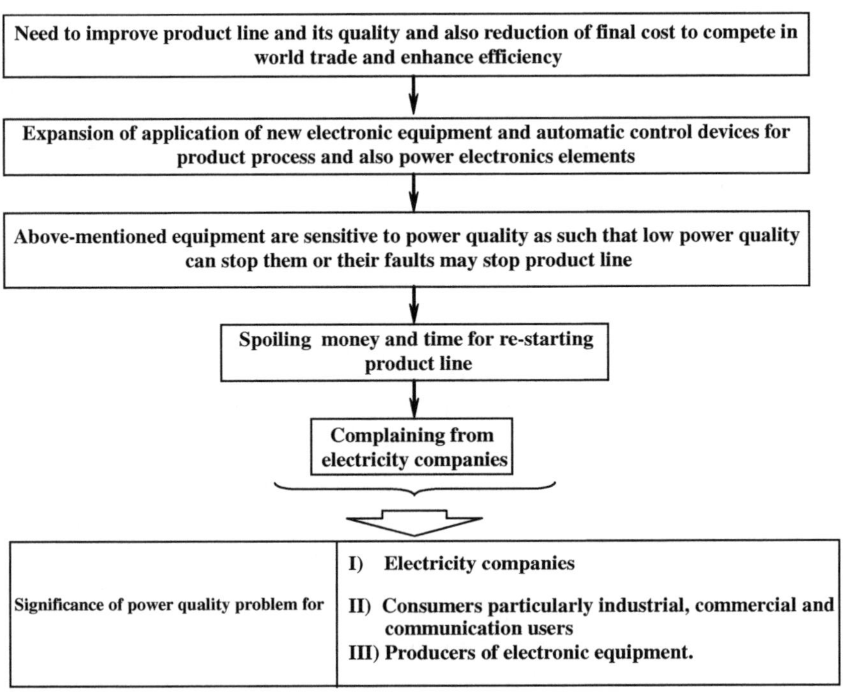

Power quality problems inflicted losses more than $27 millions in year 1991 to American companies.

(b) The above-mentioned new electronic equipment usually causes voltage disturbances (harmonics) which is a worrying problem due to the increase in the number of the equipment.
(c) Increasing the need for standardization of the power quality problem and defining performance criteria.
(d) The ever-increasing knowledge of consumers about power quality which forces the electricity companies to enhance power quality (considering competition between these companies, this is very important).
(e) Ability to measure the power quality leads to more attention to this matter.
(f) Reliability of network is increased and thus new power quality problems prove their effects (for example in a network no electricity for 2 h per day, power quality has no chance of being considered).

5.2 Power Quality Phenomena

5.2.1 Classification of Power Quality Phenomena

Any deviation from the ideal voltage or current waveforms is defined as "power quality phenomena". There are various methods to classify these phenomena [7]. A precise and correct classification of these phenomena has a considerable impact on their recognition, origin of generation and methods of measurement. IEC has classified these phenomena based on their creators. Table 5.1 presents this classification [8].

Table 5.1 IEC standard classification of power quality phenomena

Conducted low-frequency phenomena	Harmonic, inter-harmonics
	Signal systems (power line carrier)
	Voltage fluctuations
	Voltage dips and interruptions
	Voltage imbalance
	Power–frequency variations
	Induced low-frequency voltages
	DC in ac networks
Radiated low-frequency phenomena	Magnetic fields
	Electric fields
Conducted high-frequency phenomena	Induced continuous wave voltages or currents
	Unidirectional transients
	Oscillatory transients
Radiated high-frequency phenomena	Magnetic fields
	Electric fields
	Electromagnetic fields
	Continuous waves
Electrostatic discharge phenomena	–
Nuclear electromagnetic pulse	–

Table 5.2 IEEE standard classification of power quality phenomena [8]

Categories	Typical spectral content	Typical duration	Typical voltage magnitude
1.0 Transients			
1.1 Impulsive			
1.1.1 Nanosecond	5 ns rise	<50 ns	
1.1.2 Microsecond	1 μs rise	50 ns–1 ms	
1.1.3 Millisecond	0.1 ms rise	>1 ms	
1.2 Oscillatory			
1.2.1 Low frequency	<5 kHz	0.3–50 ms	0–4 pu
1.2.2 Medium frequency	5–500 kHz	20 μs	0–8 pu
1.2.3 High frequency	0.5–5 MHz	5 μs	0–4 pu
2.0 Short duration variations			
2.1 Instantaneous			
2.1.1 Sag		0.5–30 cycles	0.1–0.9 pu
2.1.2 Swell		0.5–30 cycles	1.1–1.8 pu
2.2 Momentary			
2.2.1 Interruption		0.5–30 cycles–3 s	<0.1 pu
2.2.2 Sag		30 cycles–3 s	0.1–0.9 pu
2.2.3 Swell		30 cycles–3 s	1.1–1.4 pu
2.3 Temporary			
2.3.1 Interruption		3 s–1 min	<0.1 pu
2.3.2 Sag		3 s–1 min	0.1–0.9 pu
2.3.3 Swell		3 s–1 min	1.1–1.2 pu
3.0 Long duration variations			
3.1 Interruption, sustained		>1 min	0.0 pu
3.2 Undervoltages		>1 min	0.8–0.9 pu
3.3 Overvoltages		>1 min	1.1–1.2 pu
4.0 Voltage imbalance		Steady state	0.5–2%
5.0 Waveform distortion			
5.1 DC offset		Steady state	0–0.1%
5.2 Harmonics	0–100th H	Steady state	0–20%
5.3 Interharmonics	0–6 kHz	Steady state	0–2%
5.4 Notching		Steady state	
5.5 Noise	Broad-band	Steady state	0–1%
6.0 Voltage fluctuations	<25 Hz	Intermittent	0.1–7%
7.0 Power frequency variations		<10 s	

Table 5.2 shows the electromagnetic phenomena according to IEEE [8] . This table classifies the power quality phenomena based on amplitude and duration. To recognize power quality problems and their measurement, classification in this table has a specific importance. Each phenomenon in Table 5.2 has been examined widely in many publications.

5.2.2 *Sag*

Among various phenomena of power quality, sag and also temporary interruption are the most important problems for industrial and commercial consumers and can damage them largely. These damages consist of stopping the production line and producing wastes, re-starting production line, dropping quality of produced goods and dissatisfaction of the consumers. For instance, sag in Hisu-Chu city in Taiwan (most solid-state devices factories of Taiwan are located in this city) caused damages between $100,000 and $1,000,000 [9].

Although temporary interruptions are more destructive than the sag, total damages caused by the sag is higher than that of temporary interruption because the number of sags are more. Extensive studies in USA and Canada indicate that 92% of the power quality events arise from 40 to 50% sag shorter than 2 s [10].

The main reason behind sag is the short circuit problems. A short circuit fault in a distribution feed creates a serious short term drop in the same feeder which usually happens due to the protection system operation and terminates to an interruption. However, in the adjacent feeders this fault causes short term drops which are not very serious, and this will be ended by operation of that feeder protection system.

5.3 Solutions for Power Quality Problems

Generally power quality problems are either imposed on the consumers from the network or are transferred from the consumers to the network. For instance, sag which is caused by short circuit arising from lightening is a problem that is imposed on consumers from network, while the non-linear current of the consumer generates harmonics in the network voltage, and this problem imposed by consumers on the network; of course this causes inconvenience to other consumers as well.

To prevent the problems that are created by consumers, it is necessary to enforce laws to confront the destructive effects of the loads upon power quality. However, there are three methods to decrease the damages caused by problems imposed on consumers by:

(a) Applying some modifications and strategies to the structure and maintenance of the network to reduce the faults and also their working period.
(b) Increasing tolerability of the equipment used by the consumers or making the equipment non-sensitive to power quality phenomena.
(c) Inserting interface equipment between the consumers and network to eliminate the power quality problems such that the fault in network does not stop or create problems in the consumer equipment. These equipments are power electronics devices and are employed in distribution networks and known as custom power tools.

Methods (a) and (b) are necessary but not enough and are sometimes are costly. However, nowadays, application of custom power tools is very common. Custom power tools are classified into four groups:

(a) Electronic switches such as solid-state breakers and solid-state transfer switches (STS).
(b) Series equipment such as dynamic voltage restorers (DVR) and full-electronic tap-changers.
(c) Parallel equipment such as static VAR compensator and distribution static compensator (DSTATCOM).
(d) Series–parallel equipment such as unified power quality conditioner (UPQC).

More explanations on the characteristics and performance of each of the above-mentioned groups can be found in the relevant publications [11–40].

As seen, the electronic tap-changer is proposed as a series equipment of custom power. It is of course clear that the electronic tap-changer cannot remove all power quality phenomena. For example, it cannot eliminate interruption or modify the waveform, but this equipment can reduce the most important problem of the power quality namely sag.

5.4 Using Electronic Tap-Changer for Sag Mitigation

5.4.1 Modeling Short-Circuit Fault

In Sect. 5.2.2, the short circuit fault was proposed as the major factor for the sag in the network. To check the capability limit of electronic tap-changer for compensating the voltage drop due to the short circuit faults, it is necessary to model a short circuit in an appropriate and simple way. Figure 5.1 presents the voltage distribution model used to calculate voltage drop caused by short circuits in a radial system.

In this model, Z_S is the supply impedance; Z_F is the impedance between the point of common coupling (PCC) and the short circuit point. Z_F consists of the cable impedance, short circuit impedance and also probably the transformer supplying the fault feeder. In fact, PCC is a common point that supplies the load as well as the fault feeder. In this model, the load current (before and after fault) is neglected. Therefore, no voltage drop is considered between the load and PCC. By these assumptions, following short circuit (V_{sag}), the load terminal voltage is obtained as follows:

$$V_{\text{sag}} = \frac{Z_F}{Z_S + Z_F} E \tag{5.1}$$

where E is the generator voltage. It is clear from Eq. 5.1 that a larger Z_F/Z_S leads to a lower voltage drop. This occurs when the fault feeder is supplied by a power transformer which has a relatively large impedance.

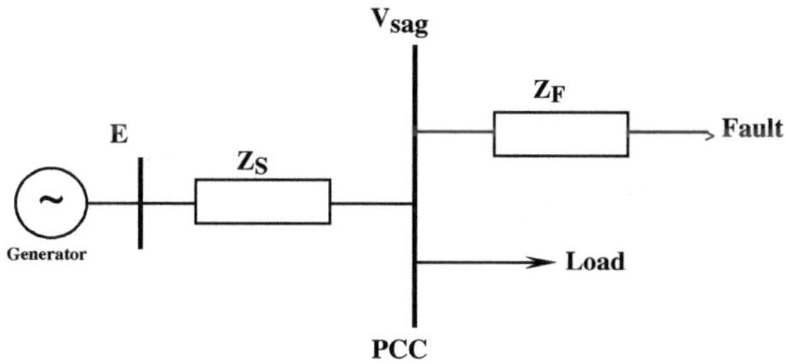

Fig. 5.1 Voltage distribution model for qualitative calculation of voltage drop due to a short circuit in a radial system

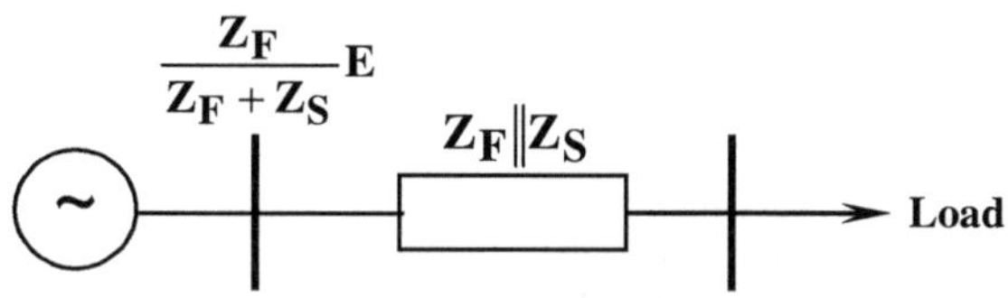

Fig. 5.2 Short circuit model seen supplying load feeder

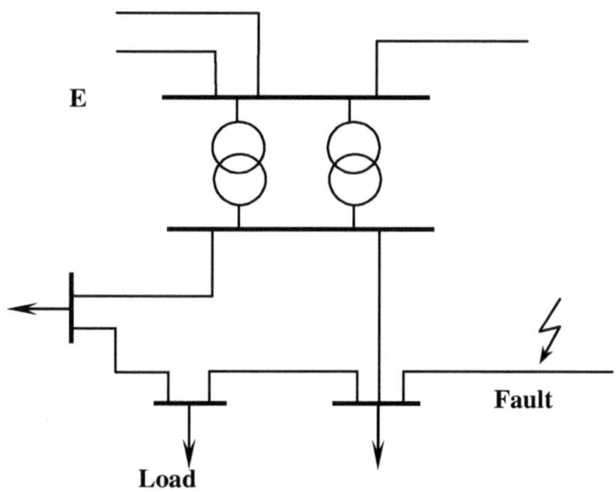

Fig. 5.3 Short circuit fault in a non-radial network

Referring to model of Fig. 5.1 and the Thevenin equivalent circuit seen from the feeder supplying the load, the short circuit fault is modeled as shown in Fig. 5.2. In the other words, a short circuit fault is modeled as the generator voltage drop equal to $\frac{Z_F}{Z_S+Z_F}E$ and decrease of the supply impedance by value of $Z_S \| Z_F$.

In non-radial systems, however complicated, short circuit in network can be modeled as a voltage drop in the amplitude of the generator voltage and a change in the supply impedance. For instance, Fig. 5.3 shows a short circuit fault in the non-radial system. Impedance model of this system has been shown in Fig. 5.4. Following short circuit, the terminal voltage of the load and impedance between

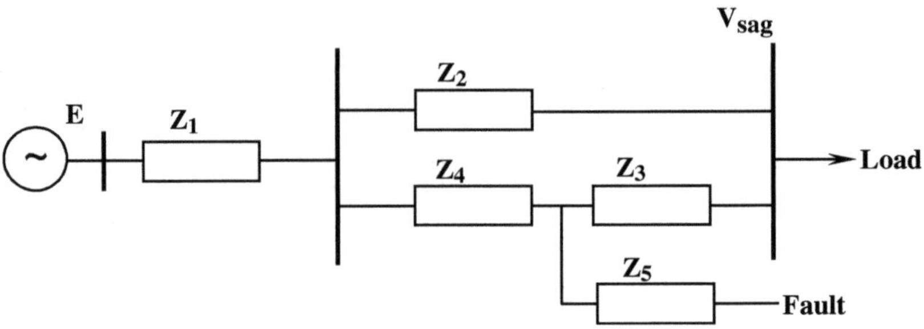

Fig. 5.4 Impedance model of Fig. 5.3

Fig. 5.5 Thevenin equivalent circuit viewed from load terminal for short circuit in network

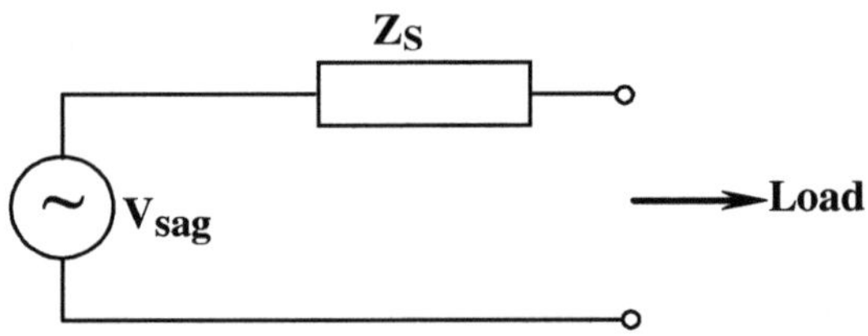

the load and supply can be written as Eqs. 5.2 and 5.3 referring to the impedance medel.

$$V_{\text{sag}} = \frac{Z_3Z_4 + Z_2Z_5 + Z_3Z_5 + Z_4Z_5}{Z_1Z_2 + Z_1Z_3 + Z_1Z_4 + Z_5Z_2 + Z_5Z_3 + Z_5Z_4 + Z_4Z_2 + Z_4Z_3}E \tag{5.2}$$

$$Z_s^{-1} = \frac{\left[\left(\frac{1}{Z_1} + \frac{1}{Z_2} + \frac{1}{Z_4}\right)\frac{1}{Z_3} + \frac{1}{Z_4} + \frac{1}{Z_2}\right]\frac{1}{Z_5} + \left[\frac{1}{Z_3Z_4} + \left(\frac{1}{Z_3} + \frac{1}{Z_4} + \frac{1}{Z_5}\right)\frac{1}{Z_2}\right]\frac{1}{Z_1}}{\left(\frac{1}{Z_1} + \frac{1}{Z_2} + \frac{1}{Z_4}\right)\left(\frac{1}{Z_3} + \frac{1}{Z_4} + \frac{1}{Z_5}\right) - \frac{1}{Z_4^2}}$$

$$Z_s = \frac{Z_1Z_2Z_5 + Z_1Z_2Z_3 + Z_2Z_4Z_5 + Z_2Z_3Z_5 + Z_2Z_3Z_4 + Z_1Z_4Z_5 + Z_1Z_3Z_5 + Z_1Z_3Z_4}{Z_1Z_2 + Z_1Z_3 + Z_1Z_4 + Z_2Z_4 + Z_2Z_5 + Z_3Z_4 + Z_3Z_5 + Z_4Z_5} \tag{5.3}$$

Referring to Eqs. 5.2 and 5.3, it is clear that $V_{\text{sag}} < E$ and $(Z_1+(Z_2||Z_3+Z_4)) > Z_S$. From the load feeder point of view, the Thevenin equivalent of the remaining network can be considered as Fig. 5.5. In Fig. 5.5 V_{sag} varies between 0 and E and its value depends on the location of the short circuit and network impedances (E is the generator voltage).

In the above discussions, it was assumed that the short circuit fault does not occur in the load feeder, but it happens in other feeders of the network. If this assumption does not exist, the short circuit activates the protection equipment of that feeder and so the fault leads to full interruption, and practically electronic tap-changer cannot help in this case.

5.4.2 Compensating Range of Electronic Tap-Changer

To estimate the compensating range, the optimal electronic tap-changer system of Fig. 5.6 is used. If the desirable voltage of the secondary is shown by V_{2n}, minimum and maximum of input voltage by $V_{P_{\min}}$ and $V_{P_{\max}}$ and assuming that electronic tap-changer is designed as such that it adjusts the output voltage for the total range of input voltage variations, then the following can be written:

$$\frac{n_{So}}{n_P} = \frac{V_{2n}}{2}\left(\frac{1}{V_{P_{\min}}} + \frac{1}{V_{P_{\max}}}\right) \tag{5.4}$$

$$\frac{n_{\text{step}}}{n_P} = \frac{V_{2n}}{48}\left(\frac{1}{V_{P_{\min}}} + \frac{1}{V_{P_{\max}}}\right) \tag{5.5}$$

For primary rated voltage V_{1n} and maximum rise of the input voltage 110%, and also voltage changes step in the secondary for changing a tap number in rated conditions of the primary 2% of voltage V_{2n}, then:

$$V_{P_{min}} = 53.5\%\ V_{1n} \tag{5.6}$$

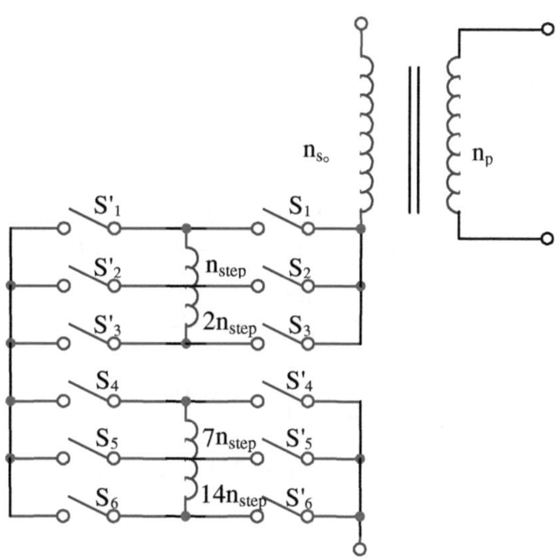

Fig. 5.6 Optimal electronic tap-changer used for estimation of compensating range

$$n_{\text{step}} = \frac{V_{2n}}{V_{1n}} \times \frac{n_P}{50} \tag{5.7}$$

$$n_{S\circ} = 1.39 n_P \times \frac{V_{2n}}{V_{1n}} \tag{5.8}$$

Due to the more destructive impact of voltage drop compared to voltage rise, efforts are made to have wider compensation range for voltage drop in an electronic tap-changer. So up to now, it is known that the proposed tap-changer is capable to compensate 53.5–110% of the primary voltage, assuming that Eqs. 5.6–5.8 are held and all series impedances (including the supply impedance, transformer impedance, switches impedance and low voltage line impedance) are equal to zero. However, if the series impedances are taken into account, then drop on these impedances will shorten the compensation range. Meanwhile, tap-changer needs a larger primary current in order to compensate the input voltage drop and in turn increases the series impedances drop.

Tap-changer compensation range for changing the load current and its angle in different networks with various supply impedance is investigated here. Figure 5.7 shows the model used for such study.

Impedance Z_{eq} is the Thevenin equivalent up to the generator. V_{sag} is the Thevenin voltage seen from the medium line voltage. Z_C is the low voltage line impedance and Z_{SW} is the impedance of the tap-changer static switches in the switch-on states. Since in any case, four switches are in series in the current path, thus their impedances are added up and gives an equal impedance Z_{SW}. Z_T(b) models the series impedance of the transformer. n_{Seq} is the secondary effective number of turns which is defined considering tap position and finally i_L is the load current and V_L is the load terminal voltage. If row number of tap in electronic tap-changer of Fig. 5.6 is called T and ratio $\frac{n_{seq}}{n_p}$ is presented by b ($b = a^{-1}$) then:

$$b = b_o + (T - 25) b_{\text{step}}; \quad T \in \{1, 2, \ldots, 49\} \tag{5.9}$$

where

$$b_{\text{step}} = \frac{n_{\text{step}}}{n_p}, \quad b_o = \frac{n_{so}}{n_p}.$$

If transformer is taken to be a distribution transformer with the Iran standard, then: $V_{n1} = 20$ kV and $V_{n2} = 400$ V, so based on Eqs. 5.7 and 5.8:

$$\begin{gathered} b_o = 1.39 \times \frac{400}{20{,}000} = 0.0278 \\ b_{\text{step}} = \frac{400}{20{,}000} \times \frac{1}{50} = 0.0004 \\ b = 0.0278 + (T - 25) \times 0.0004 \Rightarrow \begin{cases} b_{\max} = 0.0374, & T = 49 \\ b_{\min} = 0.0182, & T = 1 \end{cases} \end{gathered} \tag{5.10}$$

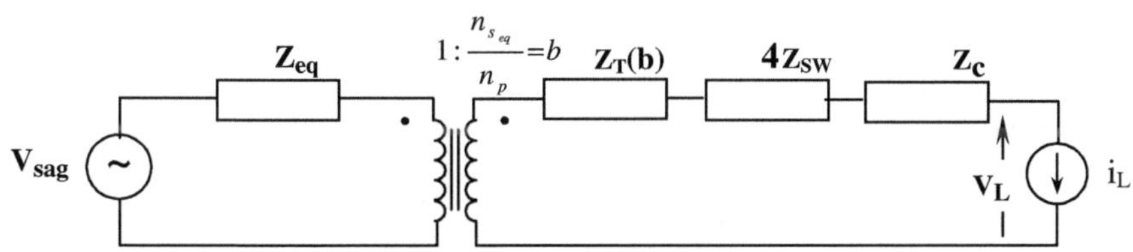

Fig. 5.7 Model used in investigation of electronic tap-changer compensation range

Considering the model presented in Fig. 5.7, the load voltage (V_L) is as follows:

$$V_L = \frac{\mathrm{b}\, V_{\mathrm{sag}}}{1 + \frac{Z_{\mathrm{eq}}\mathrm{b}^2 + Z_T(\mathrm{b}) + 4Z_{\mathrm{SW}} + Z_C}{Z_L}} \tag{5.11}$$

where Z_L is the load impedance. Eq. 5.11 is a phasor and to obtain the fixed range of Z_L value for different V_{sag} computer computations are required. However, for more profound insight to this range, it can be easily shown that Eq. 5.11 has a real answer for b if:

$$\left(\frac{V_{\mathrm{sag}}}{V_{2n}}\right)^2 - 4\frac{Z_{\mathrm{eq}}}{Z_L}\left(1 + \frac{Z_T(b) + 4Z_{\mathrm{SW}} + Z_C}{Z_L}\right) \geq 0 \tag{5.12}$$

where $\frac{Z_T(b)+4Z_{\mathrm{SW}}+Z_C}{Z_L}$ is negligible against 1, then condition (5.12) will be as follows:

$$V_{\mathrm{sag}} \geq 2\sqrt{\frac{Z_{\mathrm{eq}}}{Z_L}}V_{2n} \tag{5.13}$$

Referring to Eq. 5.13, it is seen that by rising Z_{eq} the compensation range reduces. Condition (5.13) is a necessary and sufficient condition to have real answer b. However, the b can be applied to the tap-changer if it places between $b_{\min}$ and $b_{\max}$. In other words, condition (5.13) shows a range that can fix the secondary voltage of transformer by adjusting its turns ratio, but providing this turn ratio through tap-changer depends on the limitation of tap-changer system and transformer taps.

Now in real conditions (actual impedances and not necessarily Ohmic and limitation of b between $b_{\min}$ and $b_{\max}$) the range of voltage adjustment, estimated by computer computations, is discussed. In these computations, numerical values of the parameters are as follows:

$$\begin{aligned}
Z_L &= \frac{400}{|I_L|}(\cos\alpha + j\sin\alpha) \\
|I_L|_{\max} &= 1000\,\mathrm{A} \\
Z_{\mathrm{eq}} &= 4 + j5 \\
Z_C &= 18\times10^{-3} + j8.7\times10^{-3} \\
Z_T(b) + 4Z_{\mathrm{SW}} &= 0
\end{aligned} \tag{5.14}$$

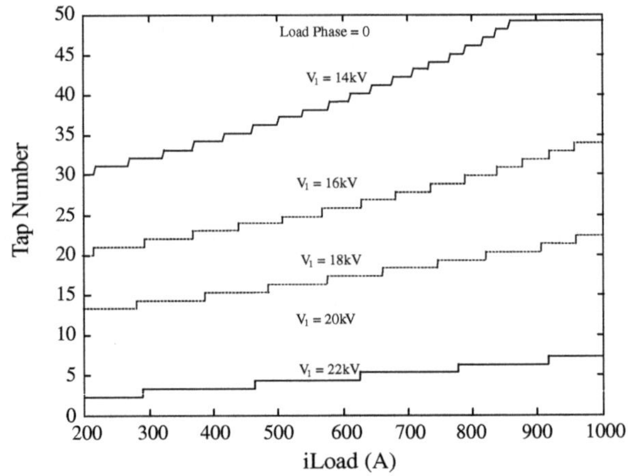

Fig. 5.8 Necessary tap number to fix V_L for variations of V_{sag} with different i_L (zero load angle)

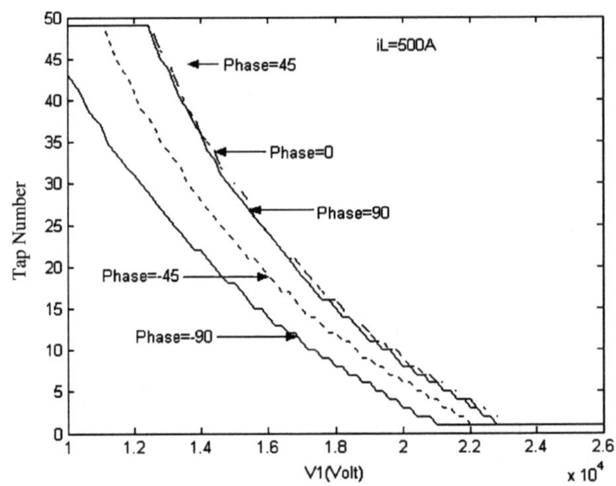

Fig. 5.9 Necessary tap number to fix V_L for variations of V_{sag} at different load angle and $i_L = 500$ A

Computations results have been shown in Figs. 5.8, 5.9, 5.10, 5.11. There are some common data in these figures; each has been plotted from a specific point of view. Figure 5.8 exhibits the necessary tap number for adjusting V_L for variations of V_{sag} in different i_L. In these curves the load impedance angle is assumed equal to zero. As seen at a constant current i_L, T decreases by rising V_{sag} and increases by reduction of V_{sag}. The maximum capacity of the system for compensating V_{sag} will be when T approaches number 49. In this case, the positions of the switches will be such that the maximum voltage is induced in the secondary of transformer. This range occurs at the load current 250 A for $V_{\mathrm{sag}} = 0.85V_{1\mathrm{n}}$, while this happens at load current 1,000 A for $V_{\mathrm{sag}} = 0.975V_{1\mathrm{n}}$.

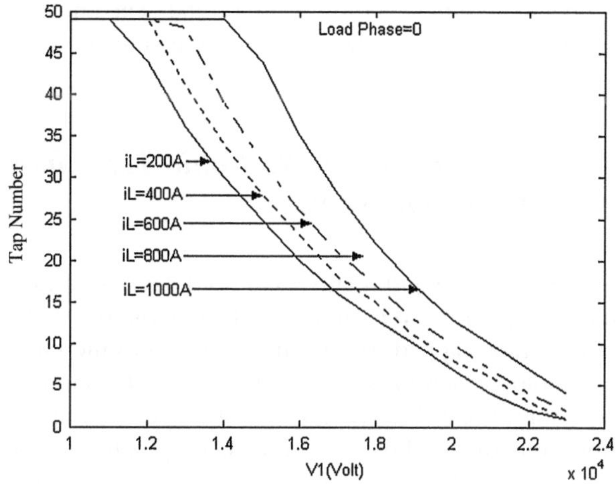

Fig. 5.10 Necessary tap number to fix V_L for variations of load current at different V_{sag} and load angle aero

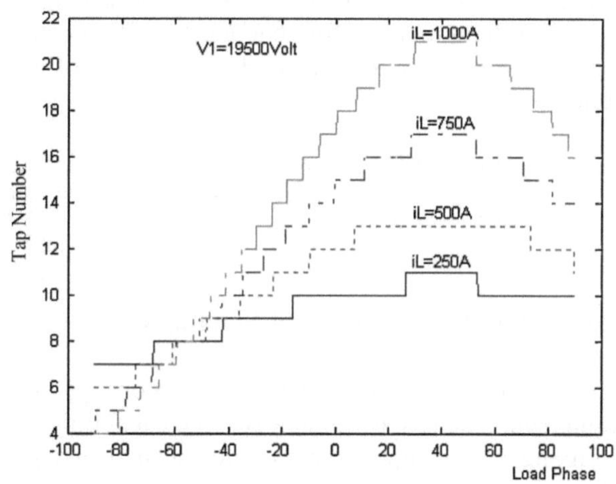

Fig. 5.11 Necessary tap number to fix V_L for variations of load angle at different i_L and $V_{sag} = 19{,}500$ V

Figure 5.9 presents the necessary value of T to fix V_L for variations of V_{sag} and different load impedance angles. In this curve, the load current is fixed at 500 A. It is seen that the range of voltage drop compensation is larger at capacitive currents and the reason is that the capacitive load contributes to voltage rising.

Figure 5.10 shows the impact of the load current variations. Certainly by rising i_L, the relevant T increases and the i_L value at which T tends to 49 depends on V_{sag}. Smaller V_{sag} leads to the lower current.

Finally Fig. 5.11 presents the variations of the load impedance angle. It is clear that for capacitive angles, necessary T for fixing V_L reduces and by rising the load angle, T also rises and at a particular angle T maximizes. If the load angle rises further, T will have a diminishing trend. The angle causing the peak T is the angle at which the load angle is equal to the sum of system impedances angle.

5.4.3 Comparison of Electronic Tap-Changer and Other Sag Mitigation Tools

So far, various equipments have been suggested to mitigate the sag. They include uninterrupted power supply (UPS), dynamic voltage restore, Ferro-resonance transformer, tapped transformer with electronic tap-changer, static transfer switch and motor-generator set. Certainly, any of the equipment have its own capabilities, advantages and disadvantages. For a specified application with its special conditions, one of the above-mentioned equipment is more economical. To mitigate the sag, a method for economical evaluation in application of the above-mentioned equipment is introduced below. It is noted that the application of the static transfer switch involves an independent feeder and thus it has not been considered in this comparison. In order to evaluate the merits of application of the above-mentioned equipment, the case in which they are not used is also considered.

The method to apply for economical comparison is estimation and comparison of the cost of each equipment. The total cost of each equipment is estimated as follows:

$$\begin{aligned}\text{Total Cost (TC)} &= \text{Purchase Price (PP)} + \text{Operating Costs (OC)} \\ &\quad + \text{Residual Cost of Sag \& Interruption (RC)}\end{aligned}$$

5.4.3.1 Purchase, Fixing and Starting Cost

Since several parameters have been proposed to determine the cost of each equipment (such as distributor, specific capabilities of equipment and complexities of fixing), it is impossible to give the real and precise cost of the equipment. However, the typical cost of these equipments can be demonstrated. Figure 5.12 exhibits a typical cost diagram of the equipment versus their power in $/kWh.

5.4.3.2 Operating Cost

This cost consists of maintenance costs, and energy costs related to the losses of the equipment. UPS battery, as an energy storage element, needs maintenance. Almost any 4 years, UPS batteries must be replaced. The cost of this replacement is about 25% of the purchasing cost of the equipment.

Motor-generator sets have moving mechanical parts and need service and maintenance. The cost of this maintenance is about 10% of the purchasing cost.

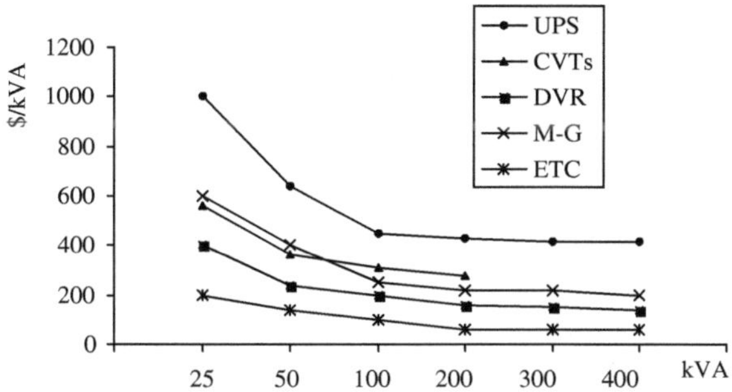

Fig. 5.12 A typical cost diagram of the equipment versus their power [40]

Other equipment are full-electronic and almost have zero maintenance cost. It is noted that DVR has no storage element.

Concerning the energy cost of the equipment losses, typical efficiency of UPS, DVR, Ferro-resonance transformer, tapped-transformer with electronic tap-changer and motor-generator set are taken to be 92, 98, 9, 84%, respectively. The cost of one kVAh energy is assumed to be $0.1. Meanwhile, OCs in the coming years is estimated now by considering the utilization factor. At present, this utilization factor is considered to be 0.8.

5.4.3.3 Remaining Phenomena Cost

Certainly each discussed equipment has its specific capabilities. Thus, they can eliminate or stop only some of sags or interruptions. Therefore, the damage cost of remaining phenomena must also be considered economically.

Figure 5.13 presents a typical density of occurrence of sags and interruptions [2]. Referring to this figure, it is clear that the total phenomena along a year is equal to 50.2. It must now be investigated how much of these phenomena can be eliminated by each of the mentioned equipment. UPS can eliminate all of these phenomena due to its unique characteristics. DVR (if has no energy storage element) cannot eliminate the interruptions longer than two, three cycles and as shown in Fig. 5.13, these phenomena occurs 3.1 times per year, so DVR eliminates about 94% of the phenomena.

Ferro-resonance transformers can eliminate about 78% of the phenomena which is only about 60–70% of sags as seen in Fig. 5.13.

Tapped-transformer with the electronic tap-changer can to eliminate up to 50% of the sags considering the calculations in Sect. 5.2.2, so it eliminates about 85% of the phenomena. Finally, motor-generator set eliminates about 98% of the phenomena. Of course, the total number of the phenomena in different places differs, but in many cases density of various sag distributions (amplitude and duration) is as shown in Fig. 5.13. Therefore, percentage of the cases that are eliminated by each of the discussed equipment is extendable to any case and location.

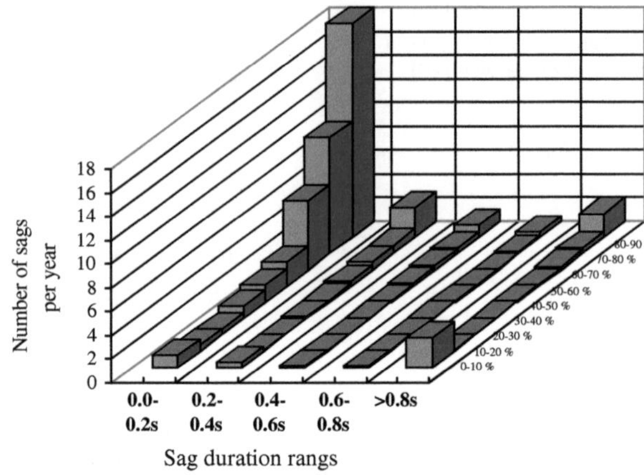

Fig. 5.13 Typical density of occurrence of various sags and interrupts [2]

Another problem is the estimation of the damage caused by each phenomenon. Since these damages depend totally on the particular conditions of the consumers and equipment, there is no a specific number, but it can be estimated between 50 and 400 \$/kVA.

Thus, the cost relevant to the damage due to the remaining phenomena is as follows:

$$\begin{aligned}&\text{Annual cost of damage due to remaining phenomena}\\&\text{per year for nth year transferred to present} = \frac{n \times \text{Load} \times \text{Cost}}{(i+1)^n}\end{aligned} \quad (5.15)$$

where N is the average remaining phenomena per year in that place, load is the consumed power of the consumer (in kVA), Cost is the cost of one phenomenon (in \$/kVA), n is the gain factor and n is the year. Now, an example is given to compare the discussed equipment economically for compensating the sag. In any other case, the same method can be applied in order to select the most economical equipment.

Example A 50 kVA load is considered and assumed that its sensitive cost to any phenomenon is 50 \$/kVA, if this load confronts 25 phenomena every year, application of which equipment is more economical?

Table 5.3 summarizes the total cost of the first up to fifth year of the example. According to this table, the cost of the first year due to the electronic tap-changer is lower than that of the other equipment, from second year upward the total cost is minimal for DVR. It is clear that if the distribution density of the occurrence of sag types is such that the amplitude of the major sags becomes larger than 50%, full-electronic tap-changer can be considered as most economic equipment.

Table 5.3 Comparison of costs of equipment mitigating sag for given example

Equipment name	PP (k$)	RC for 1st year	RC for 1st year	Maintenance cost for 1st year	Total annual cost (k$)				
					1st	2nd	3rd	4th	5th
UPS	34	0	2.152	1.168	37.293	40.342	43.154	45.789	48.192
DVR	14	3.750	0	0.292	18.042	21.784	25.236	28.47	31.421
CVT	17	13.75	0	3.650	34.400	50.511	37.265	79.292	91.994
ETC	6	9.375	0	0.146	15.521	24.336	32.468	40.085	47.035
M-G	20	1.250	2	2.336	25.586	30.613	35.53	40	44
None	0	62.5	0	0	62.5	120.37	75.173	75.223	269.375

References

1. Billinton R, Allan RN (1984) Power system reliability in perspective. IEE J Electro Power 30(3):231–236
2. Bollen MHJ (2000) Understanding power quality problems. IEEE Press Series on Power Engineering
3. Koval DO, Bocancea RA, Yao K, Hughes MB (1998) Canadian national power quality survey: frequency and duration of voltage sags and surges at industrial sites. IEEE Trans Ind Appl 34(5):904–910
4. Dugon RC, Granaghan MF, Beaty HW (1996) Electronics power system quality. McGraw-Hill, New York
5. Reynders K (>>>>) Power quality, LABOR ELEC Doc
6. Bonatto BD, Mertens Jr EA, da Silva ES, Dioo LFS (2002) Power quality assessment at sensitive loads, T & D 2002, Mar 2002
7. Hughes BM, Chan JS, Koval DO (1993) Distribution customer power quality experience. IEEE Trans Ind Appl 29(6):1204–1211
8. IEEE Recommended Practice For Monitoring Electric Power Quality (1995) IEEE Std. 1159
9. Brumsickle WW, Schneider RS, Luckjiff GA, Divan DM (2001) Dynamic sag correctors: cost effective industrial power line conditioning. IEEE Trans Ind Appl 37(1):212–217
10. Bingham RP (1998) Sags and swells, Dranetz-MBI (on-line). Available http://www.dranetz-bmi.com/pdf/sags-swells.pdf
11. Custom Power-State of the Art, Reported by CIGRE, July 2000
12. Sannino A, Miller MG, Bollen MHJ (2000) Overview of voltage sag mitigation, proceedings of IEEE PES Winter 2000 Meeting, Jan 2000, pp 2872–2878
13. Song YH, Johns AT (1999) Flexible AC transmission system (FACTS). Institution of Electrical Engineers Series 30, London
14. Stump MD, Keane GJ, Leong FKS (1998) The role of custom power products in enhancing power quality at industrial facilities, IEEE international conference on energy management and power delivery (EMPD'98), 3–5 Mar, vol 2, pp 507–517
15. Hingorani NG (1995) Introducing custom power, IEEE spectrum, June 1995, pp 41–47
16. Osborne MM, Kitchin RH, Ryan HM (1995) Custom power technology in distribution systems-an overview. IEE symposium on reliability and security of distribution systems, 1995, pp 10/1–10/11
17. Arnold R (2001) Solution to the power quality problem. Power Eng J 15:65–73
18. Sannino A (2001) Economic evaluation of power quality solutions in deregulated environment. Euro Conf Risk Manage Power Syst

19. Nilsson S (1999) Special application considerations for custom power systems. IEEE PES Winter Meeting 31 Jan–4 Feb, vol 2, pp 1127–1131
20. Taylor GA (1993) Power quality hardware solutions for distribution systems: custom power, North Eastern center power section symposium on the reliability, security and quality of distribution systems, 5 Apr, Durham, UK, pp 11/1–11/9
21. Sundaram A, Mehta H, Tahiliani V (1994) Custom power EPRI's response to power quality issues, South Con 94, pp 398–403
22. Vilathgamuwa M, Perera AADR, Choi SS, Tseng KJ (1999) Control of energy optimized dynamic voltage restorer, the 25th annual conference of the IEEE industrial electronics society (IECON '99), San Jone, CA, USA, Nov 1999, pp 873–878
23. Lee S, Choi J, Hang H (2000) Power quality enhancement in distribution line using series compensator. IEEE Power Con, pp 1595–1600
24. Middlekauff SW, Collins ER (1998) System and customer impact: considerations for series custom power devices. IEEE Trans Power Deliv 13(1):278–282
25. Cheng PT, Huang CC, Pan CC, Bhattacharya S (2002) Design and implementation of a series voltage sag compensator under practical utility conditions. 16th annual IEEE applied power electronics conference and exposition, Dalls, TX, USA, 10–14 Mar, vol 2, APEC, pp 1061–1067
26. Dahler P, Affolter R (2000) Requirements and solutions for dynamic voltage restorer. A case study, IEEE PES Winter Meeting, 23–25 Jan, vol 4, pp 2881–2885
27. Awad H, Svensson J (2002) Compensation of unbalanced voltage dips using vector-controlled static compensator with LC filter. 37th IEEE IAS annual meeting conference record, vol 2, pp 904–910
28. Wunderlin T, Amhof D, Dahler P, Guning H (1998) Power supply quality improvement with a dynamic voltage restorer (DVR), international conference on energy management and power delivery (EMPD'98), 3–5 Mar, vol 2, pp 518–525
29. Sannino A, Svensson J (2001) Application of converter-based series devices for voltage sag mitigation to induction motor load, IEEE power tech conference, Porto Portugal, vol 2, pp 6
30. Sannino A, Svensson J (2002) Static series compensator for voltage sag mitigation supplying nonlinear loads. IEEE Power Eng Soc Winter Meet 2:1147–1152
31. Raithmayr W, Dahler P, Eichler M, Lochner G (1998) Customer reliability improvement with a DVR or a DUPS. Power World 98, Santa Clara, Nov 10, pp 1–10
32. Wang P, Jenkins N, Bollen MHJ (1998) Experimental investigation of voltage sag mitigation by an advanced static VAR compensator. IEEE Trans Power Deliv 13(4):1461–1467
33. Jung SY, Kim TH, Moon S, Han M (2002) Analysis and control of DSTATCOM for a line voltage regulation. IEEE Power Eng Soc Winter Meet 729–734
34. Clouston JR, Gurney JH (1999) Field demonstration of a distribution static compensator used to mitigate voltage flicker. IEEE PES summer meeting, Santiago, 31 Jan–4 Feb, vol 2, pp 1138–1141
35. Santoso S, Zavadil R, Folts D (2001) Modeling and analysis of a 1.7 Mva SMES-based sag protector, IPST' 01, Brazil
36. Graovac D, Katic V, Rufer A (2000) Power quality compensation using universal power quality conditioning system. IEEE Power Eng Rev 58–60
37. Faiz J, Siahkolah B (2004) Effect of solid-state on-load distribution tap-changer on power quality enhancement. Int J Eng, IR Iran 17:143–156
38. Faiz J, Siahkolah B (2005) Sag mitigation by an electronic tap-changer: specifications and comparisons with other custom power tools. IEE Proc, Pt. C 152(5):697–704
39. Zouiti M, Saadate S, Lombard X, Poumarede C, Levillain C (1998) Electronic based equipment for flicker mitigation. IEEE 1CHQP 98, Athens, Oct 1998, pp 1182–1187
40. Degeneff RC, Barss R, Raedy S (2000) Reducing the effect of sags and monentary interruptions: a total owning cost prospective, IEEE 9th International Conference on Harmonics and Quality of Power, Oct 2000, vol 2, pp 397–403

Chapter 6
Practical Implementation and Experimental Results

6.1 Technical Specifications and Block Diagram of Prototype

To verify the design of the power section and also control of the electronic tap-changer experimentally a low-power prototype tap-changer has been built and tested. Technical specifications of this electronic tap-changer set are as follows:

Number of phases	1
Output fixed voltage (rms)	220 V
Input rated voltage (rms)	220 V
Number of voltage steps	48
Step voltage in input rated conditions (rms)	2 V
Maximum output current (rms)	20 A
Power of set	5 kW

Block diagram of the prototype electronic tap-changer has been shown in Fig. 6.1 [1].

Hardware of the blocks are described in Sect. 6.2. Details of microprocessor control block are given in Sect. 6.3. Finally experimental results are introduced and discussed in Sect. 6.4.

6.2 Hardware

In this section the hardware of each part of the prototype electronic tap-changer set is introduced and described briefly.

6.2.1 Tapped-Transformer

Figure 6.2 exhibits the schematic circuit used in the prototype electronic tap-changer. In this figure the transformer secondary consists of a main winding with

J. Faiz and B. Siahkolah, *Electronic Tap-changer for Distribution Transformers*, Power Systems, DOI: 10.1007/978-3-642-19911-0_6,

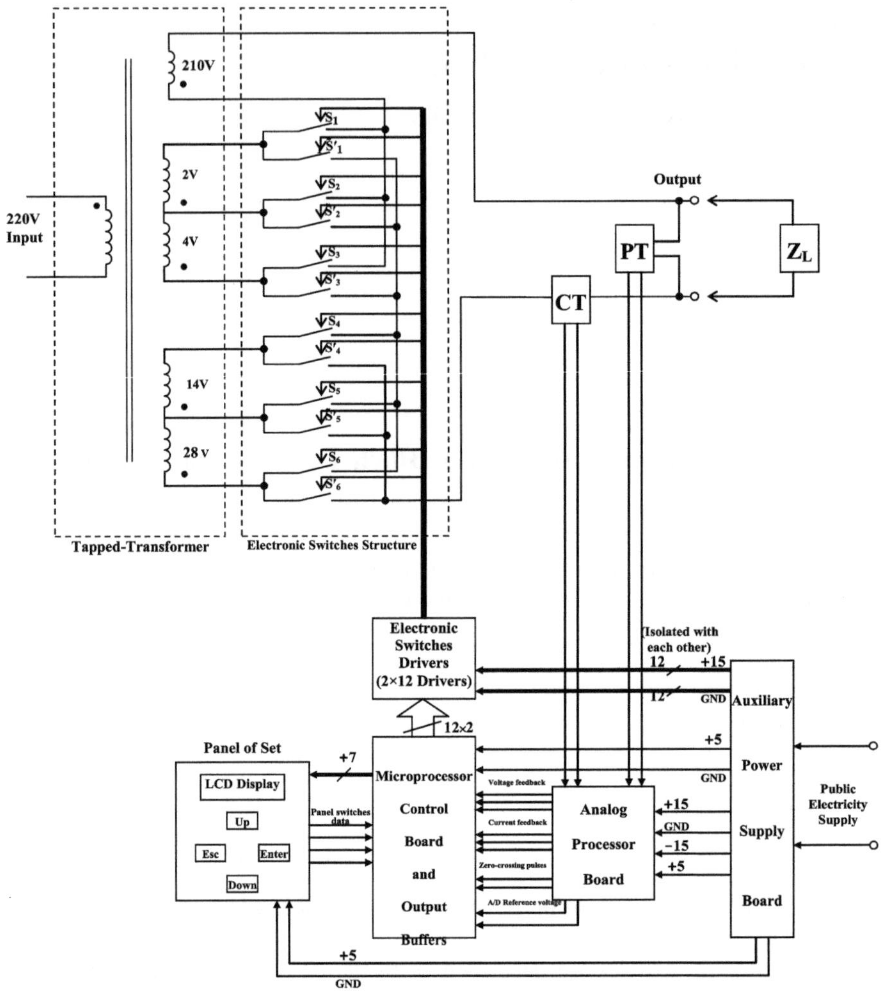

Fig. 6.1 Block diagram of prototype electronic tap-changer

the rated voltage of 210 V, two tapped-windings each has three taps and the rated voltages of "2 V, 4 V" and "14 V, 28 V". The rated secondary voltage means a voltage that is induced when the primary voltage is 220 V. As mentioned in Chap. 2, this configuration has an optimal topology and its six taps can provide 49 different cases for output voltage. If this transformer is connected to the appropriate tap-changer it can transform input voltae of 187 V up to 298 V into the regulated secondary voltage of 220 ± 1 V based on the turns ratio, where the series impedances are neglected. The size of wires of all windings has been selected based on the rated output current (20 A).

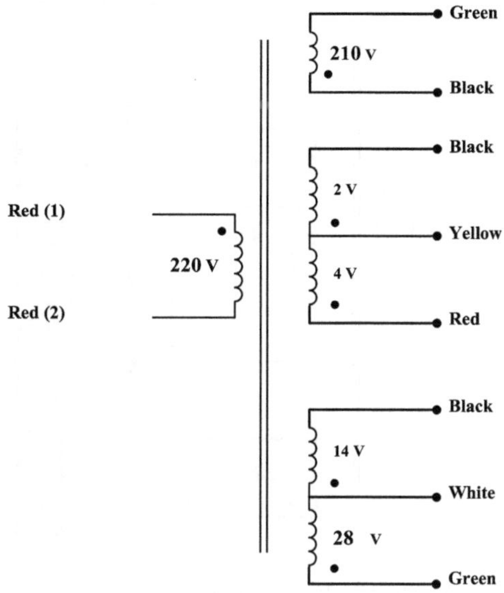

Fig. 6.2 Schematic circuit of transformer used in electronic tap-changer

6.2.2 Electronic Tap-Changer Switches

Figure 6.3 shows the schematic circuit of electronic tap-changer switches and their connection to the tapped-transformer. As mentioned in Chap. 3, to realize bi-directional solid-state switches two uni-directional solid-state switches (MOSFET or IGBT) are used. Meanwhile, the uni-directional switches are connected such that their sources (in MOSFET) or emitters (in IGBT) are connected to each other, so there is no need to have two isolated power supplies. If in each one of the following four groups of switches: (S_1, S_2, S_3), (S_1', S_2', S_3'), (S_4, S_5, S_6), (S_4', S_5', S_6') one switch is on at all time (even at switching time), then the maximum voltage on each of electronic tap-changer switches are as follows (it is noted that this maximum voltage is true if the input voltage does not go over 300 V):

$$
\begin{aligned}
&\text{Maximum voltage} = S_1, S_1', S_3, S_3' = 6\sqrt{2} \times \frac{300}{220} = 11.6\,\text{V} \\
&\text{Maximum voltage} = S_2, S_2' = 4\sqrt{2} \times \frac{300}{220} = 7.7\,\text{V} \\
&\text{Maximum voltage} = S_4, S_4', S_6, S_6' = 42\sqrt{2} \times \frac{300}{220} = 81\,\text{V} \\
&\text{Maximum voltage} = S_5, S_5' = 28\sqrt{2} \times \frac{300}{220} = 54\,\text{V}
\end{aligned}
\tag{6.1}
$$

Switches are selected based on the maximum voltage given by Eq. 6.1 and the current of each switch. The maximum current of all switches can be written as

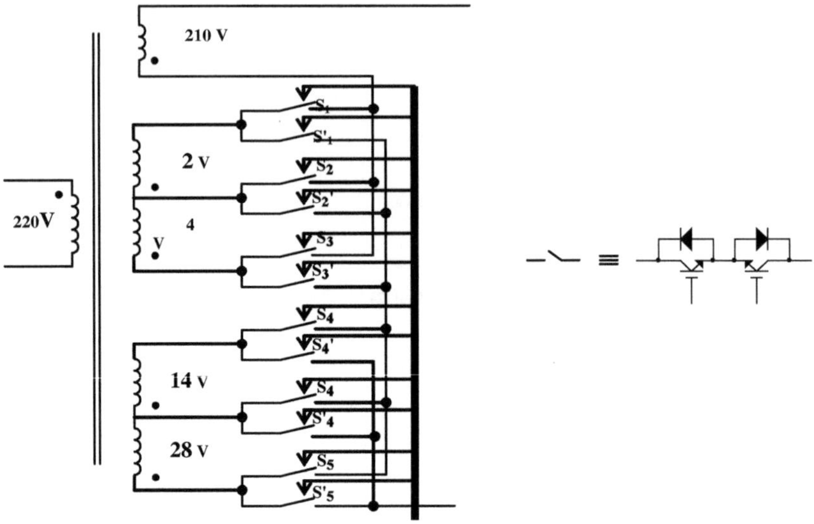

Fig. 6.3 Schematic circuit of electronic tap-changer and their connections to tapped-transformer

follows (suppose the load current does not exceed 20 A for this maximum current):

$$\text{Maximum current of each switched-on uni-directional switch} = 20\sqrt{2} = 28.3\ \text{A}$$

$$\text{Maximum average current of each switched-on uni-directional switch}$$

$$(\text{without diode}) = \frac{20\sqrt{2}}{2\pi} = 4.5\ \text{A}$$

$$\text{Maximum average current of each switched-on; uni-directional switch}$$

$$(\text{with diode}) = \frac{20\sqrt{2}}{\pi} = 9\ \text{A}$$

Table 6.1 presents the positions of switches for realization of 49 different cases in the output. Data of this table is stored as a look-up table in the control board. When the tap number changes, the corresponding position of the switches is extracted from the table. In this table V_{out} is the output voltage of the tap-changer for input voltage of 220 V.

6.2.3 Drivers of Solid-State Switches

Figure 6.4 presents the schematic circuit of each solid-state switch. Opto-coupler is used to isolate the switch-on and switch-off command signals of the switches

Table 6.1 States of one-by-one of switches corresponding to tap number

No	Switch states												V_{out} − 210 (V)
	S_1	S_1'	S_2	S_2'	S_3	S_3'	S_4	S_4'	S_5	S_5'	S_6	S_6'	
1	0	1	0	0	1	0	0	1	0	0	1	0	−48
2	0	0	0	1	1	0	0	1	0	0	1	0	−46
3	0	1	1	0	0	0	0	1	0	0	1	0	−44
4	1	1	0	0	0	0	0	1	0	0	1	0	−42
5	1	0	0	1	0	0	0	1	0	0	1	0	−40
6	0	0	1	0	0	1	0	1	0	0	1	0	−38
7	1	0	0	0	0	1	0	1	0	0	1	0	−36
8	0	1	0	0	1	0	0	0	0	1	1	0	−34
9	0	0	0	1	1	0	0	0	0	1	1	0	−32
10	0	1	1	0	0	0	0	0	0	1	1	0	−30
11	1	1	0	0	0	0	0	0	0	1	1	0	−28
12	1	0	0	1	0	0	0	0	0	1	1	0	−26
13	0	0	1	0	0	1	0	0	0	1	1	0	−24
14	1	0	0	0	0	1	0	0	0	1	1	0	−22
15	0	1	0	0	1	0	0	1	1	0	0	0	−20
16	0	0	0	1	1	0	0	1	1	0	0	0	−18
17	0	1	1	0	0	0	0	1	1	0	0	0	−16
18	1	1	0	0	0	0	0	1	1	0	0	0	−14
19	1	0	0	1	0	0	0	1	1	0	0	0	−12
20	0	0	1	0	0	1	0	1	1	0	0	0	−10
21	1	0	0	0	0	1	0	1	1	0	0	0	−8
22	0	1	0	0	1	0	1	1	0	0	0	0	−6
23	0	0	0	1	1	0	1	1	0	0	0	0	−4
24	0	1	1	0	0	0	1	1	0	0	0	0	−2
25	1	1	0	0	0	0	1	1	0	0	0	0	0
26	1	0	0	1	0	0	1	1	0	0	0	0	2
27	0	0	1	0	0	1	1	1	0	0	0	0	4
28	1	0	0	0	0	1	1	1	0	0	0	0	6
29	0	1	0	0	1	0	1	0	0	1	0	0	8
30	0	0	0	1	1	0	1	0	0	1	0	0	10
31	0	1	1	0	0	0	1	0	0	1	0	0	12
32	1	1	0	0	0	0	1	0	0	1	0	0	14
33	1	0	0	1	0	0	1	0	0	1	0	0	16
34	0	0	1	0	0	1	1	0	0	1	0	0	18
35	1	0	0	0	0	1	1	0	0	1	0	0	20
36	0	1	0	0	1	0	0	0	1	0	0	1	22
37	0	0	0	1	1	0	0	0	1	0	0	1	24
38	0	1	1	0	0	0	0	0	1	0	0	1	26
39	1	1	0	0	0	0	0	0	1	0	0	1	28
40	1	0	0	1	0	0	0	0	1	0	0	1	30
41	0	0	1	0	0	1	0	0	1	0	0	1	32
42	1	0	0	0	0	1	0	0	1	0	0	1	34
43	0	1	0	0	1	0	1	0	0	0	0	1	36

(continued)

Table 6.1 (continued)

No	Switch states												V_{out} – 210 (V)
	S_1	S_1'	S_2	S_2'	S_3	S_3'	S_4	S_4'	S_5	S_5'	S_6	S_6'	
44	0	0	0	1	1	0	1	0	0	0	0	1	38
45	0	1	1	0	0	0	1	0	0	0	0	1	40
46	1	1	0	0	0	0	1	0	0	0	0	1	42
47	1	0	0	1	0	0	1	0	0	0	0	1	44
48	0	0	1	0	0	1	1	0	0	0	0	1	46
49	1	0	0	0	0	1	1	0	0	0	0	1	48

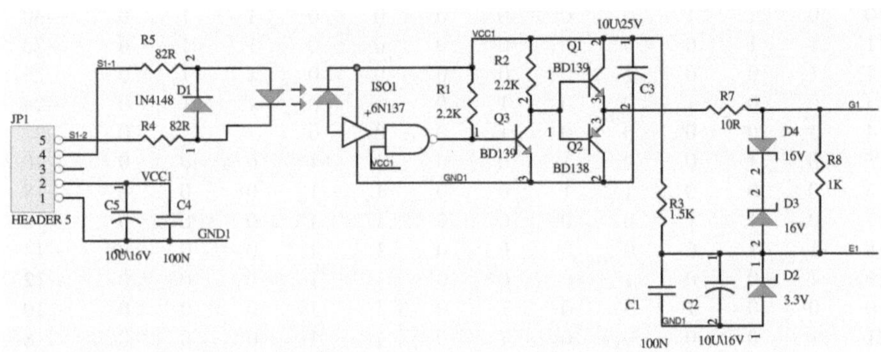

Fig. 6.4 Schematic circuit of solid-state switches of switches drivers

from the power section. After passing the command signal from opto-coupler, its current is amplified by the single-stage push–pull amplifier and is then applied to the switch.

As mentioned in Sect. 6.2.2, since the source (or emitter) of both switches is connected to each other, an isolated power supply is sufficient for the driver circuit of these switches. The required isolated power supplies for drivers are produced by auxiliary power supplies blocks.

6.2.4 Analog Processor Board

The responsibility of the analog board is to receive analog feedback signals (voltage and current of load from PT and CT) and then filtering, amplifying and rectifying them and finally delivering to the microprocessor control board.

To reduce the quantizer error in the microprocessor board processing, three different gains in the analog processing board are considered for each voltage and current signal. Meanwhile, this board generates the required reference voltage for analog/digital converter located in the microprocessor processing board. In addition, the analog processing board generates two square wave signals, by zero-

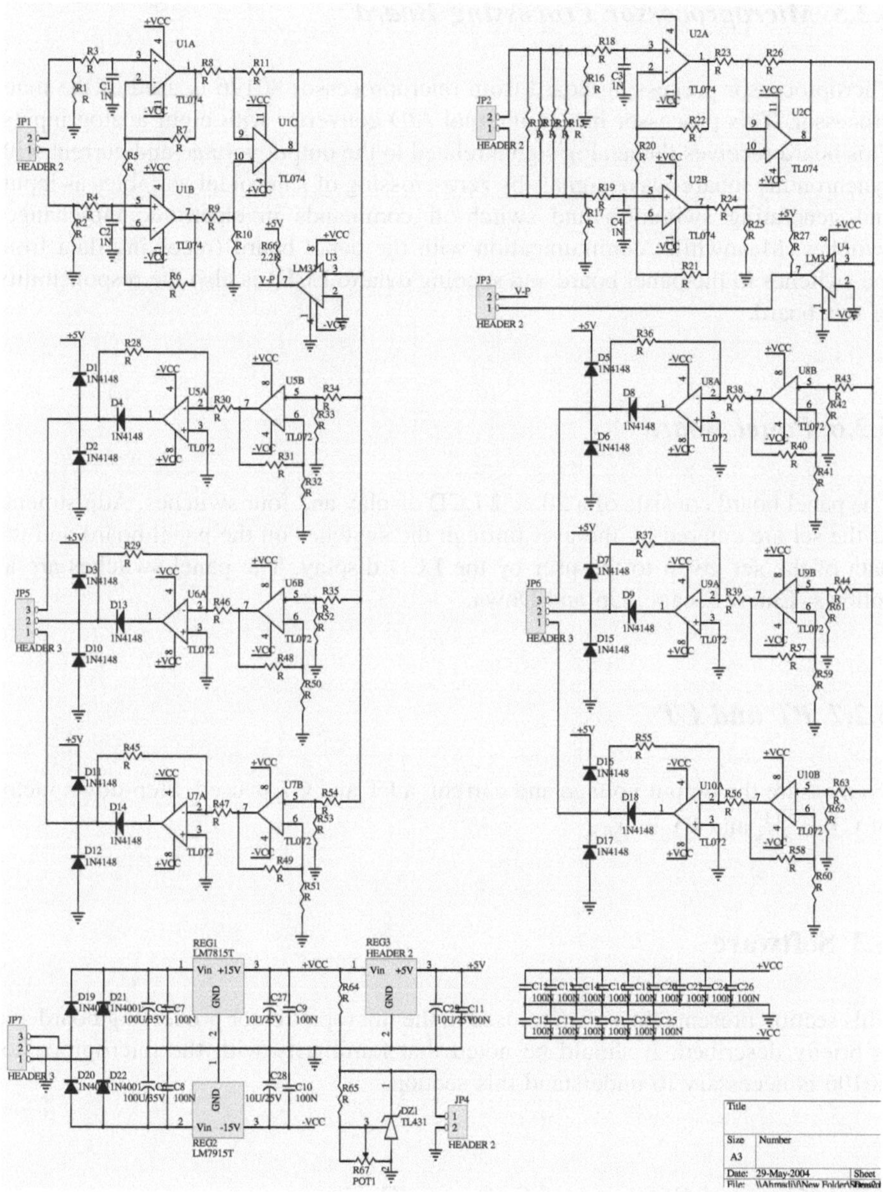

Fig. 6.5 Schematic circuit of analog processor

crossing of each voltage and current signal in a synchronous order. These square-wave signals, called digital voltage signal and digital current signal, are used in the microprocessor processing board. Figure 6.5 shows the schematic circuit of the analog processing board.

6.2.5 Microprocessor Processing Board

Microprocessor processing board from microprocessor 80196 is used as the main processor. This processor has an internal A/D converter with eight analog inputs. This board receives the analog signal related to the output voltage and current with synchronous square-wave signals by zero-crossing of sinusoidal variables as input, and generating switch-on and switch-off commands in electronic tap-changer switches. Meanwhile, communication with the panel board (receiving data from the switches in the panel board and sending data to LCD) is also the responsibility of this board.

6.2.6 Panel Board

The panel board consists of a 20×2 LCD display and four switches. Adjustments of the set are entered by the user through the switches on the panel board and the data of the set given to the user by the LCD display. The panel switches are as follows: Enter, Escape, Up and Down.

6.2.7 PT and CT

To measure the output voltage and current, a PT and CT is used. Step-down factor of CT is $\frac{0.1}{8.75}$ and PT is $\frac{5}{220}$.

6.3 Software

This section presents the software used in the microprocessor processing board and is briefly described. It should be noted that familiarity with the microprocessor 80196 is necessary to understand this section.

6.3.1 General Strategy of Software Design

Figure 6.6 presents the block diagram of the electronic tap-changer from a control point of view. The designed software in the microprocessor processing board totally fulfills the control part of this block diagram.

In order to convert the load voltage and current into the useable signals in the control system, a square-wave voltage in phase with the load voltage and a

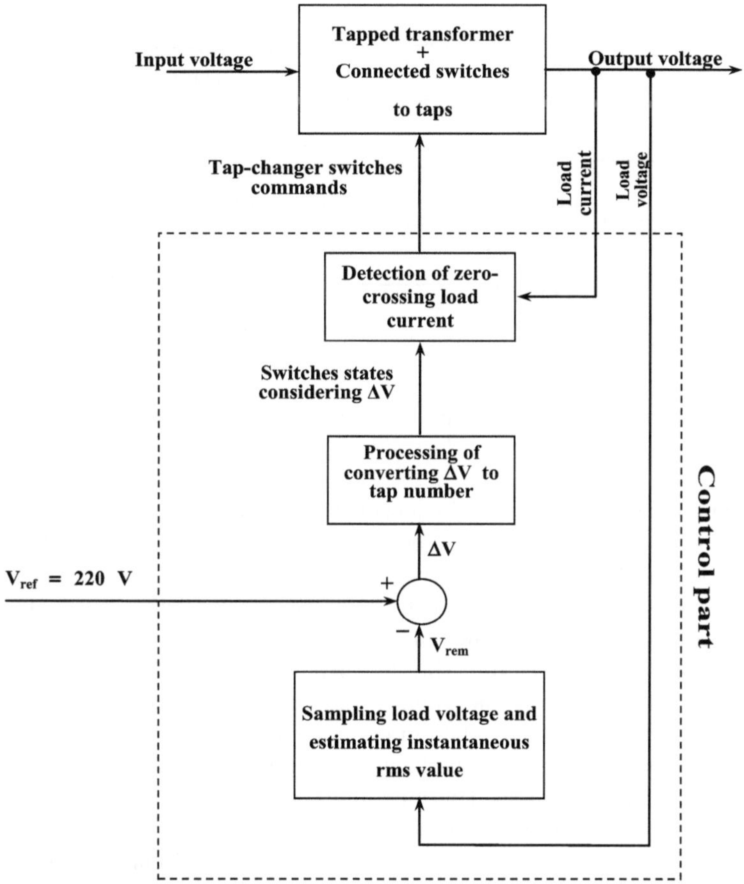

Fig. 6.6 Block diagram of electronic tap-changer from control point of view

square-wave current in phase with the load current is generated by full-wave three rectified voltages and currents in phase with the load current (Fig. 6.7), using the analog processor board. Relation of load voltage and load current with the above-mentioned signals is specified by the waveforms given in Fig. 6.8.

The reason for generating three rectified signals from the voltage (or current) is to cover a wider range of the voltage with an appropriate precision; for example if one is to cover a load voltage signal up to the maximum $350\sqrt{2}$ V, a step-down factor of $\frac{350\sqrt{2}}{5} = 99$ is used, while the maximum load voltage over most times is around the rated value $220\sqrt{2}$. So, the input signal A/D will have amplitude equal to $\frac{350\sqrt{2}}{99} = 3.14$; however, the amplitude of the input voltage of A/D converter is permissible 5 V. This is the reason to use three step-down factors; one transfers $350\sqrt{2}$ V into 4.5 V, the next one transfers $300\sqrt{2}$ V into 4.5 V and finally the last one transfers $350\sqrt{2}$ V into 4.5 V. Thus in reading the A/D channels if the

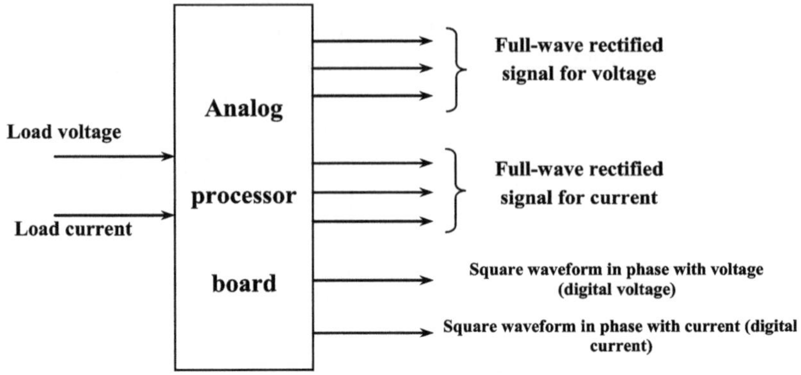

Fig. 6.7 Inputs and outputs of analog processor board

voltage amplitude of one channel exceeds 4.5 V, sampling will be done from the next channel in order to be able keep the accuracy as well as to read the higher voltages. This is also true for the current.

The reason for generating digital signals in phase with the voltage and current is that by rectifying full voltage and current waveforms, the zero-crossing type data of these signals demolishes (crossing negative to positive or positive to negative values). This data exists in the voltage and current signals and is given to the microprocessor. Meanwhile, other important characteristics of the voltage and current digital signals, which is perhaps more important than the first one, is that the edges of these two signals determine the zero-crossing of the load voltage and current, and this is a useful information. Particularly, zero-crossing current is vital for switching at the zero current. Therefore, voltage and current digital signals contain the type and zero-crossing locations of the load voltage and current.

The general strategy of designing the software is based on the following basis:

(a) Use of the digital voltage edge for:

  (a1) evaluation of an half period of the load voltage and consequently estimation of the load voltage frequency (f);
  (a2) frequency control.
  (a3) starting regular sampling of the voltage and current along an half period over the time interval calculated as follows:

$$t_{\text{samp}} = \frac{\frac{T}{2}}{\frac{n_{\text{samp}}}{2}} \tag{6.3}$$

where $T/2$ is the distance of two successive zero-crossing voltage, t_{samp} is the time between two successive sampling from input analog signals and n_{samp} is the number of samples during a complete cycle of the input voltage.

  (a4) Evaluation of the load power factor

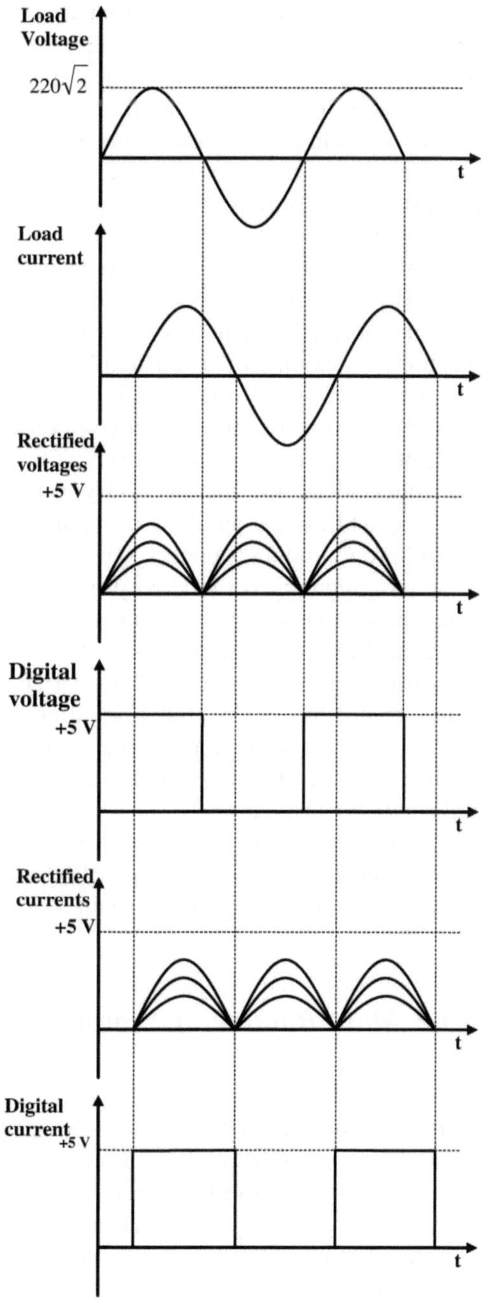

Fig. 6.8 Typical waveforms of input and output of analog processor board

(b) Use of the digital current edge for:

(b1) Evaluation of the load power factor.
(b2) Applying switches command to realize zero-current switching.

By applying the above-mentioned items, the control system has been fully applied. By coming any voltage edge, the half period time, which is now finished, is estimated, then t_{samp} is evaluated by Eq. 6.3. Following t_{samp} estimation, the block that is responsible for the regular sampling from voltage and current is begun. The sampling block calculates the instantaneous rms value of voltage (V_{rms}), evaluates rms current value (I_{rms}), estimates ΔV from V_{rms} and applies the process of transforming ΔV to the tap number. At the current edge, the load power factor estimated and the switch commands are applied.

In practice, HIS block is used to detect the digital voltage and current edges and block HSO is utilized to program the regular sampling (starting A/D over regular time intervals). Evaluation of the instantaneous rms value, ΔV and process of ΔV transformation to the tap number are done at the location of every voltage sample and interrupt routine A/D. The voltage digital signal is the basis of the system operation. Coming and edge of this signal is the indication of ending a half-cycle of voltage and beginning the next half-cycle. Block HIS detects the edges. At the edges locations, the necessary calculations are carried out and required, sampling and estimating along new half-cycle are programmed. So, practically the control programming is undertaken by the interruption routine HIS (HSIH). One of these programming is the HSO programming for starting A/D along a half cycle over appropriate intervals. HSO does the sampling control in parallel with the program. For every sample taken by A/D the interruption occurs at the end of transformation. At interrupt routine of A/D, estimation of the instantaneous rms value, evaluation of ΔV, transformation process of ΔV into the tap number are done. Upon the detection of the current edge, HIS block applies the switch commands corresponding to the calculated tap number in routine A/D to the drivers of the switches.

Referring to the above explanations, it is clear that the control system operates by the interrupt routine HIS, block HSO and interrupt routine A/D. So, all tasks are carried out in parallel with the main software program.

6.3.2 Main Routine of Program

1. Figure 6.9 shows the main program chart. Routine PON makes the system ready to start the control. Transforming the time intervals to 8 $\times$ state time and storing them in the predefined rooms in RAM, preparing LCD, estimating some parameters and storing them in RAM (to let these parameters for control process) and preparing some internal blocks of the microprocessor (such as A/D, timer 2, timer 1, HSO) to serve the control, are the tasks done in routine PON. The details of these items are given in Sect. 6.3.2.1.

Following the implementation of PON, implementing the program until the system resets again will be placed in routine MBODY. In this routine, subroutine CONTROLRESET is called by studying the bit of the resent state (CRSTSB)

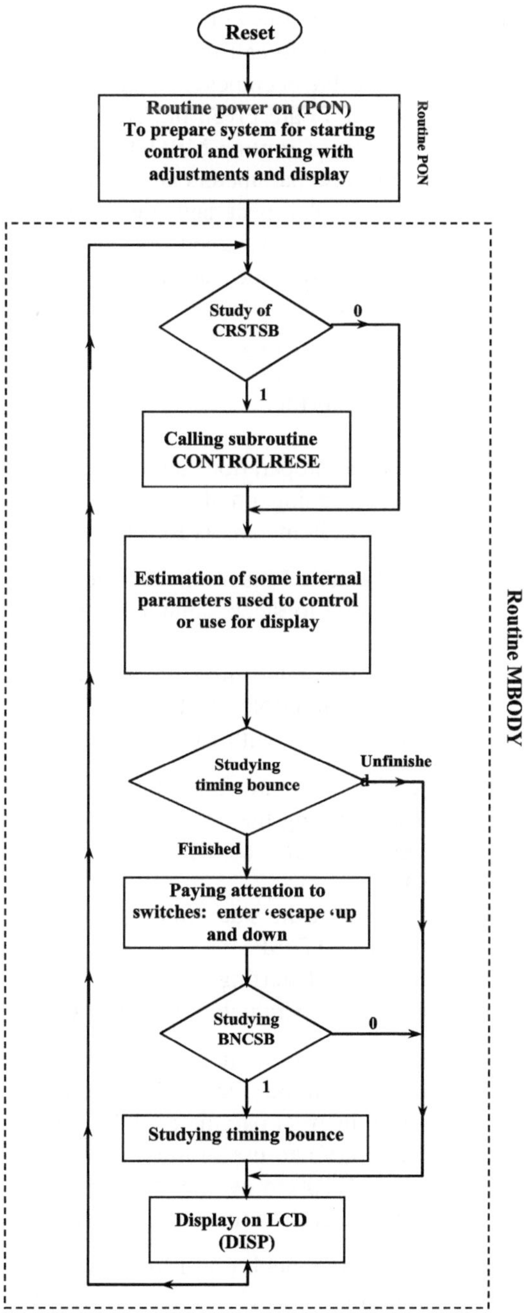

Fig. 6.9 Flow-chart of main routine of program

Responsibility of this subroutine is resetting and starting the control system, and the most important part of it is starting block HIS for consideration of the edges of the digital current and voltage signals. According to the control logic, if it is necessary to begin the control process from the start, the state bit CRSTSB becomes 1 in order to call this subroutine. For instance, when the system is reset the above state bit becomes 1 at the end of routine PON. Details of subroutine (CNTROLRESET) are given in Sect. 6.3.2.2.

The next part of routine MBODY is assigned for evaluation of some parameters. Some of these parameters are estimated in order to show them on the display; some have control aspect and are used in the control process. The remaining routine MBODY are assigned considering the switches and the display on the LCD; in other words, they are for adjustments and working with menu and displaying on LCD.

6.3.2.1 PON Routine

Calculations and preparations are carried out in this routine; they are done once at the time of switching on the system (or any time that the system is reset as software). These items are as follows:

1. To clear "enable" bit of all interruptions;
2. To give value to stack pointer for utilizing second 256 bytes RAM (SP ← 200H);
3. Estimating the state time number equivalent to bouncing time of switches and storing them in TBNC in RAM;
4. Preparing to work with LCD.
    - Estimating the state time number equivalent with different times of adjustment and storing them in TADJS, TADJM and TADJF.
    - Assigning value to MEN and TEXT (MEN = MENFI and TEXT = TEXTF1).
    - Calculation of the state time number equivalent to display time on LCD and storing it in TDISP.
    - Re-setting and starting.
    - Defining new parameters for LCD (Δ, $\surd$, φ, ▼, ▲, ◀, ▶).
5. Setting the system in an automatic tap control state (MANTCH = 01).
6. Calculating minimum manual tap adjustment (TAPMIN = −KMAX).
7. Estimating internal parameters KDV and KDV2 and assigning the initial value to DV and DV1 to use in low-pass filter.
8. Assigning INT00 to overhead of timer 1.
9. Preparing timer 2 as time base of block HSO:
    - Supplying the clock pulse from the inside (ICO3.0 = 0, ICO.7 = X).
    - Programming the clock pulse speed for unit increment in any 8statetime (ICO2.0 = 0).
    - Programming for counter-up counting (IOC2.1 = 0).
    - Inactivating internal reset (IOC0.5 = X, ICO0.3 = 0).

- Programming counter-up overhead of timer 2 in passing FFFF to 0000 (ICO2.5 = 0).
- Interrupting timer 2 accessibility to interruption INT00 (IOCT.3 = 0).

10. Preparing A/D for working with analog signals and transforming correctly.

- Estimating the content of register AD-TIME:

$$\mathrm{SAM} = \frac{T_{\mathrm{SAM}} \times F_{\mathrm{OSC}} - 2}{8}, \quad \mathrm{CONV} = \frac{T_{\mathrm{CONV}} \times F_{\mathrm{OSC}} - 3}{2 \times 10} - 1$$

- Programming A/D to use AD-TIME (IOC2.4 = X IOC2.3 = 1)
- Programming pin P0.7 in order that it does not operate as external interrupt pin.

11. Initial preparing of block HSO to operate as center of sampling start.

- Inactivating pins HSO.4 and HSO.5 as output HSO (IOC1.6 = 0, IOC1.4 = 0).
- Activating events lock in CAM (IOC2.6 = 1).

12. Putting the state bit CRSTSB equal to 1.

6.3.2.2 CONTROLRESET Subroutine

This subroutine resets and starts the control operation of microprocessors (resetting timer 1, starting block HIS and etc.). This subroutine has the bit state CRSTSB, the subroutine is called if this bit is 1. Thus, where it is necessary to reset the control operations in the program, this bit becomes 1. The following items are performed in sequence:

1. Disabling all interruptions.
2. Resetting the position of panel board switches and their bouncing (TFBNC = 0, KEYS).
3. Resetting displaying time on LCD (TFDISP = 0).
4. Estimating the ideal half period from ILFREQ and storing it in HPTIM.
5. Storing the rated power factor in COSFI.
6. Resetting tap control and applying command of switches equivalent to T = 0 to switches (TAPNO, PATT, SWPN).
7. Resetting the sampling from voltage and current (SV2SAMP = 0. SI2SAMP = 0), resetting storing voltage samples VSAMP bank ($V_{rms} = 0$, $I_{rms} = 0$) and setting channel 0 for voltage (VCH = 0) and channel 4 for current (CCH = 4).
8. Programming A/D for starting by HSO and 10 bit conversion (AD-COMMAND = 00H).

9. Preparing timer 1 for timing time intervals and also timing half period (registers TIMI reset, assigning value to bits HPMSXSB, TIOVSB and activating interrupt of overhead of timer 1).
10. Zeroing a half period number (NPNO = 0).
11. Initial assigning to the load type ICLSB = 1).
12. Preparing and starting block HSI to consider current and voltage digital signals.

   - Inactivating HIS.0 (voltage digital dignal) and HIS.1 (current digital signal).
   - Reflushing FIFO.
   - Determining HSI.0 and HIS.1 modes to consider both falling and rising edges (HIS-MODE = 0001111).
   - Activating INT02 to generate interrupt HSIH when register hold of HIS is filled (IOC1.7 = 0, INTMASK.2-1).
   - Activating interrupt INT14 to generate interrupt when FIFO is filled (INTMASK1.6 = 1).
   - Starting block HIS through activating HIS.0 and HIS.1.

13. Activating the "enable" bit for all interrupts.

6.3.3 Implementation of Control Logic (HIS, HSO, A/D, Timer 1 and Timer 2)

Interruption routine HIS, HSO block, interrupt routine A/D and timers 1 and 2 as the time base of HIS and HSO are implemented the basis of control logic. Among the above-mentioned blocks, HIS plays a more basic role. This block controls all cases by examining the edges of the digital voltage and current. It starts the sampling from voltage and current (undertaken by HSO and A/D). Among two digital voltage and current inputs, the digital voltage is more important and frequency estimation (half cycle) and starting HSO and sampling occur at the edges of this signal (edges of digital current are examined only for calculation of power factor and application of switches commands). Figure 6.10 shows the operation of the control system at the instant of resetting the system based on the digital voltage.

It is noted that HSIHV is implemented at the voltage edges and HSIHV at the current edges. Block HSO has no definite routine but is implemented at appropriate times according to HSIHV programming and AD (consisting of ADV and ADC) is implemented after each A/D conversion.

6.3.3.1 Flow-Chart of Interrupt Routine HIS (HSIH)

Figure 6.11 presents the flow-chart of HIS interruption routine. Meanwhile, HIS from timer 1 is used as a time base, so interruption routine of timer 1 is also shown in Fig. 6.12. It is noted that the zero-crossing of voltage and current digital signals

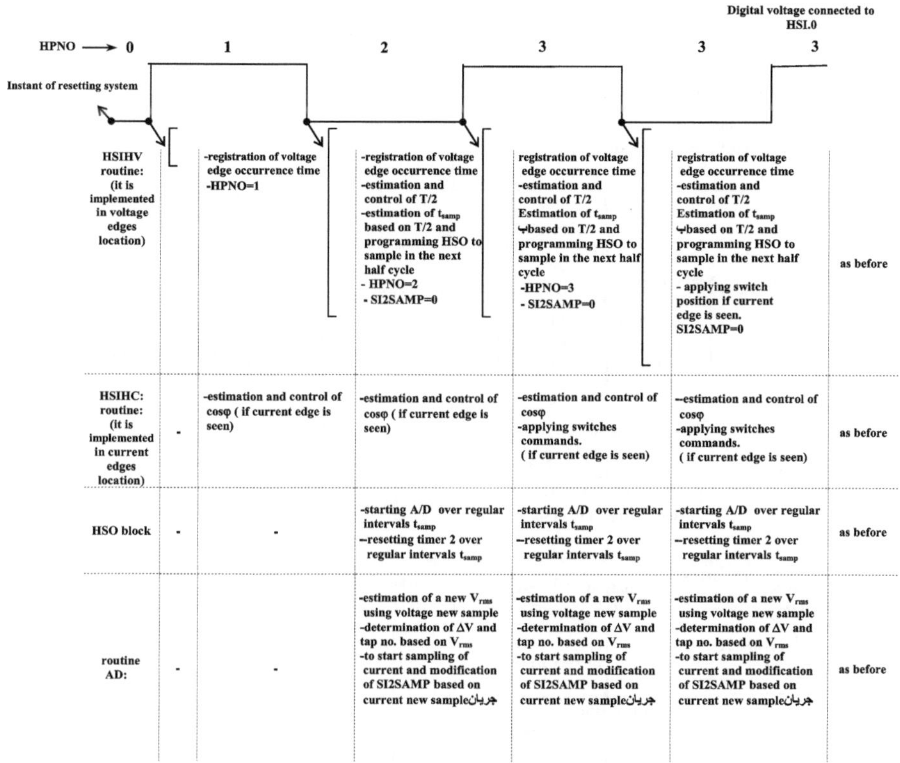

Fig. 6.10 Explanation of control system operation from resetting time of system based on voltage digital signal

are detected by block HIS. This block uses timer 1 as a time base. In order to control and evaluate the distance between edges correctly, state bits T1OVSB and HPMAXSB are defined. In fact, TIOVSB is the complementary of timer 1 value. If timer 1 overheads, this bit becomes 1 until the value of this timer is used. It is known that it is already overhead.

When an edge comes, in interrupt routine HIS (USIH):

(a) Interrupt HIS disables such that it does not generate interrupt for coming new interrupt during the processing the present edge.
(b) Its event and time is stored. Event information is in HIS-STATUS, time information is in HIS-TIME and bit state is in TIOVSB that are stored in HSISTI, HSITIMI, T1OVISB respectively.
(c) If the edge is the voltage edge or edge of voltage/current:

- Pin HIS.0 (digital voltage) is inactivated, so far another voltage event was not basically observed.

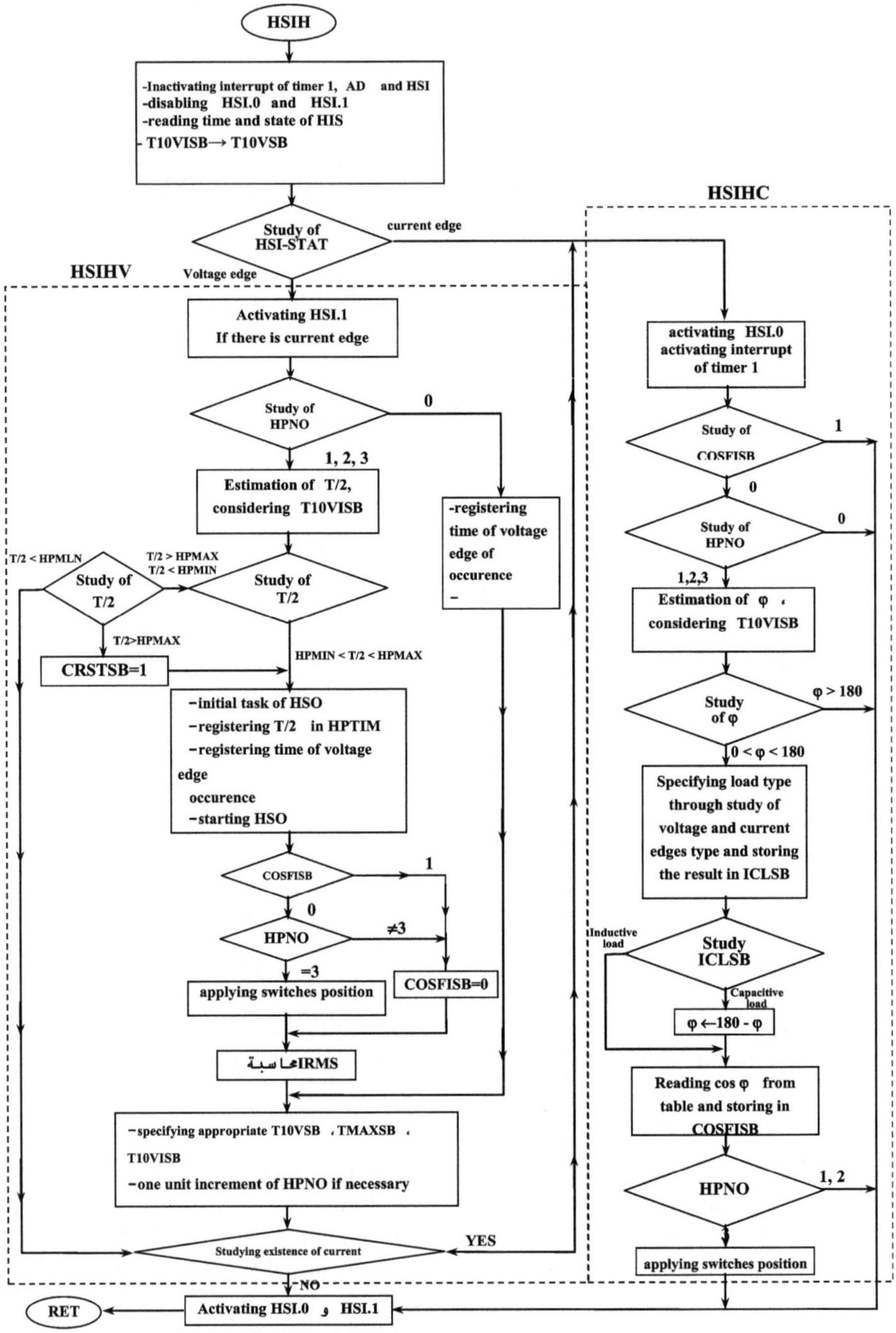

Fig. 6.11 Flow-chart of interrupt routine HIS (HSIH)

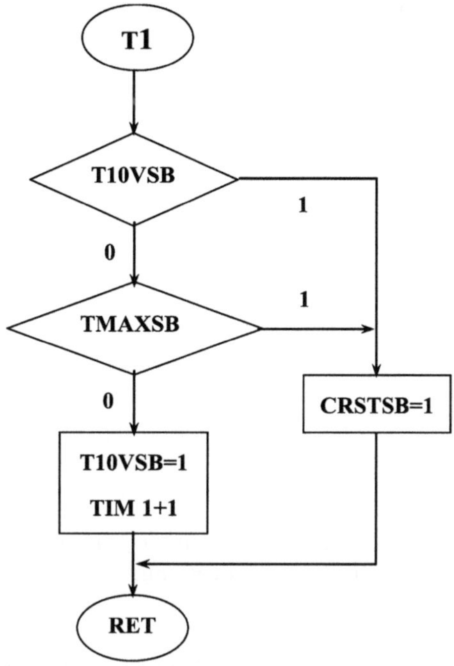

Fig. 6.12 Interrupt routine of timer 1

- HIS.1 pin (digital current) is activated only in the voltage edge case and is inactivated in the voltage/current edge case.
- Interrupt of timer 1 is inactivated. The reason is that if the edge occurs close to the overhead time of timer 1 and probably overhead of timer is half-cycle error, overhead of the timer during edge processing which does not generate interruption. Then two cases may be occurred:
- *Edge acceptance*: T1OVSB = 0, T1OVISB = 0, determination of HPMAX based on VZCT value and activating timer 1 interrupt.
- *Edge non-acceptance*: no change of TIOVSB, no change of HPMAX and activating timer 1 interrupt.

(d) If the edge is the current edge:

Pin HIS.1 (digital current) is inactivated, so far another voltage event was not basically observed and processing on this edge ends. An important point concerning block HSI is noted here, it is possible to fill FIFO due to successive edges. To pay attention to this case, FIFO FULL interrupt in control system is also activated and its interrupt routine is called HSIF. In this interrupt routine, six filled rooms (of seven rooms) are called and discarded and only information of the 7th event (last one) is preserved for processing (Fig. 6.13). Although their information is discarded, there is at least a control on the discarded information. Of course,

Fig. 6.13 Flow-chart of interrupt routine HSIF

other strategies can be employed for discarding FIFO information and discards the information after processing.

6.3.3.2 Flow-Chart of Interrupt Routine of A/D (AD)

Figure 6.14 shows the flow-chart of interrupt routine A/D.

Basically A/D is ready to start and sample the voltage by block HSO. When this task is performed at the location of each sample, program implementation is transferred to the interrupt routine AD (section ADV). When the sample voltage processing is finished in ADV, routine AD is ready to sample the current and start. This time, implementation of the program is transferred to AD (section ADC) at the end of sampling. At the end of the current sampling process, A/D is ready to start by HSO and sample the voltage at the location of the next sample (t_{samp} second latter).

6.3.3.3 Flow-Chart of Interrupt Routine of Timer 2 (T2OV)A/D (AD)

Timer 2 interruption is activated in interrupt routine HIS (HSIH). In this case, timer 2 is used as time base for HSO, and during HSO starting for sampling in HSIH, in addition to starting A/D for HSO, resetting timer 2 is also defined, as such that after resetting of timer 2 it always rests at any T_{samp}; and this is the reason that the overhead of timer 2 will not occur at all. So, if program implementation transfers to T2OV, it means that timer 2 does not set on-time and this in turn will announce the error of block HSO of the microprocessor. In interrupt routine of T2OV, system is reset.

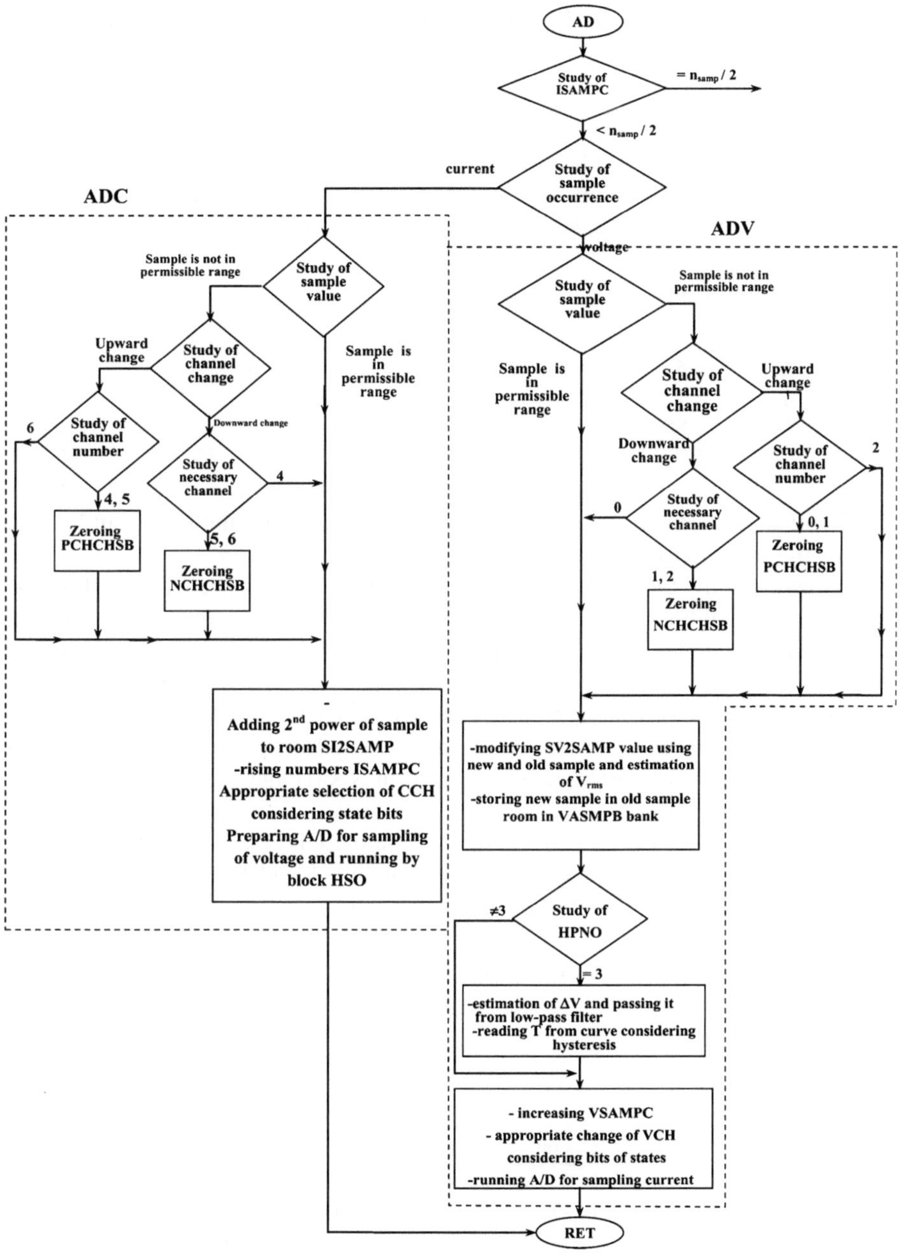

Fig. 6.14 A/d Interrupt routine (AD)

6.3.4 Display and Regulations of System

6.3.4.1 LCD

The menu of the system has been built on a 20 × 20 LCD. Data is exchanged with LCD as 4 bits and thus 7 bits are assigned for the control. Therefore, a latch of output (8 bits) with address 8000H controls the LCD by assigning the following bits:

Latch Output Lines:

7	6	5	4	3	2	1	0
↓	↓	↓	↓	↓	↓	↓	↓
X	E	$\overline{SW}$	RS	DB7	DB6	DB5	DB4

Of course, it is also possible to exchange 8 bits data with the LCD, but in order to save the output consumptions, 4 bits data exchange is used.

To begin using the LCD, resetting and initializing is needed. At this end, subroutines LCDI4 and LCDR4 are developed. In these subroutines, the resetting and initializing cycles of the LCD are performed according to the catalogue. On the other hand, use of LCD involves sending data and instruction with LCD. To do this, subroutines LCDWD4 and LCDWi4 have been developed (based on the LCD catalogue). To send data or instruction, it is enough to place its 6 bits value in registrar called AXB (which already defined in RAM) and the above subroutines are called.

Other subroutines developed to use LCD are LCD14 AND LCDL24, These subroutines put the cursor at the beginning of line 1 and line 2 of LCD respectively, so as to make it possible to write the arbitrary characters on the proposed line.

Therefore, the six subroutines proposed in Table 6.2 have been developed to work with the LCD.

Table 6.2 Six developed subroutines for LCD operation

Operation	Subroutine name
LCD R4	Performing LCD reset cycle
LCD 14	Initializing LCD cycle
LCD W14	Sending instruction yo LCD (AXB must be already filled)
LCD WD4	Sending data to LCD (AXB must be already filled)
LCD L14	Transferring cursor to beginning of 1st line of LCD
LCD L24	Transferring cursor to beginning of 2nd line of LCD

Table 6.3 List of non-exist characters in LCD internal table and defined for LCD at initializing stage

Character	Assigned code
►	00H
►	01H
▲	02H
▼	03H
Φ	04H
√	05H
Δ	P6H

The LCD is initialized only once at the beginning of the system switch-on (PON routine). It means that first LCDR4 and then LCD14 are called. This is sufficient for initialization. Henceforth, in any part of the program, different jobs can be performed by the other four subroutines.

Here, some specific characteristics which do not exist in Table 6.2 are also used. For this reason, these characteristics have been defined for LCD at the initializing stage. These characteristics have been introduced in Tables 6.3 and 6.4.

6.3.4.2 Menu of Display Page

There is a main page on the menu that is shown on the LCD immediately after is switched-on the system. Access to the next pages by enter key and previous pages by escape key is possible. If there are more than two lines written on the LCD in one page of data menu, upper and lower lines are visible by up and down keys.

As seen, all pages of the menu have been classified into 5 groups. In this classification, all pages that are ramified from a specific page are placed in one group. Then one number is assigned to every page of the menu. The most left digit of this number is the number of the page group and the next digits are the number of page group in its group. For the pages that are alone in its group, row number 1 is considered (See Table 6.5).

Assigning a row number to a page in a particular group is totally arbitrary. Number of rows begins from 1 and number is assigned to different pages in upward to downward sequence.

To call group 3 pages, row numbers from 1 to 16 are continuous and all at once row number 20 is assigned to a page that comes from page 01. This is also totally arbitrary and is done due to the different types of that page in comparison with other pages of group 3.

It should be noted that other criterion that is considered in grouping the page, is that all pages in which the number adjustment is done are placed in group 3. In other groups, display and displacement in the menu is merely carried out. This is exactly the reason that from manual tap adjust in 0 group goes all at once to group 3 in manual control state, in order to change the tap. So group 3 is also called adjust group.

Table 6.4 System display menu pages

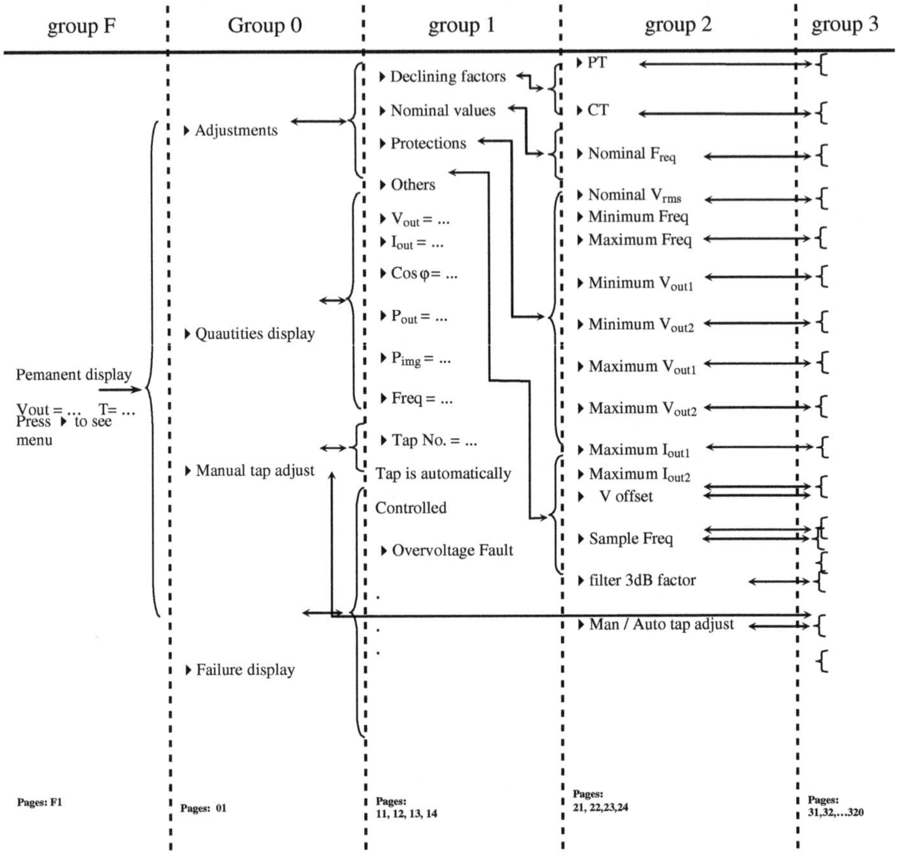

Table 6.5 Assigning number to every page of menu

Group	Existing pages
F	F1
0	01
1	11, 12, 13, 14
2	21, 22, 23, 24
3	31, 32, 33, 34, 35, 36, 37, 38, 39 310, 311, 312, 313, 314, 315, 316 320

6.3.4.3 Keys

To put the menu in the practice and to do adjustment in the group 3 of menu, four keys, namely enter' escape' up' and down' are considered. Operation of any key depends on the menu page. This is the reason that operation of each key in

Table 6.6 Keys codes and corresponding operation to each code

Enter code	Operation
0	–
1	Displacement in menu with no condition
2	Displacement in menu with given condition
3	Save number task
4	Item save task
Escape code	**Operation**
0	–
1	Displacement in menu
Up and down code	**Operation**
0	–
1	Displacement between page items
2	Timing adjustments of number
3	Adjustments of number as unit by unit with save task

different pages are identified and coded. Then in table MEN related to each page, its operating code is given to the microcontroller based on the operation of each key. Table 6.6 gives the key codes and the corresponding operation with each code. Table 6.7 shows that what is the code of each key in each page of the table.

The keys positions are examined in MBODY1. Examination of the keys is done by priority. The priority of the keys is from upper priority enter, escape, up and down respectively. When a key with higher priority is active, examination of the lower priority is ignored; while a key is active or newly activated, it continues to be active considering a code that the key has in that page of the menu. Figures 6.15, 6.16 and 6.17 show the procedure for this.

The following points must be noted about flow-charts of Figs. 6.15 and 6.16:

1. In the examination of keys, the meaning of 00, 01, 10 and 11 has been introduced in Table 6.8.
2. ADJNUM is a word that has been defined in RAM. When we enter a new page by passing enter key, if code of the enter in this page be equal to 3 (it means that adjustments are done by down/up keys) a numerical value that is desired to be adjusted, is placed in ADJNUM until adjusts keys down/up, ADJNUM (not numerical value itself) and then the adjusted value (ADJUNM) is stored in the main number location by reactivating the enter key.
3. This is always the case that if a key be active (either now or before), after giving an appropriate service to the key itself, examination of the next keys is ignored because of priority of the keys.

The following points must be noted about flow-charts of Figs. 6.17 and 6.18:

1. In the examination of the keys, meaning of 00, 01, 10 and 11 have been introduced in Table 6.8.
2. Refer to flow charts (Figs. 6.15, 6.16) for explanation of ADJNUM.

Table 6.7 Code of any key in each page

Page number	Enter Code	Escape code	Up code	Down code
F1	1	0	0	0
01	2	1	1	1
11	1	1	1	1
12	0	1	1	1
13	0	1	0	0
14	0	1	1	1
21	1	1	1	1
22	1	1	1	1
23	1	1	1	1
24	1	1	1	1
31	3	1	2	2
32	3	1	2	2
33	3	1	2	2
34	3	1	2	2
35	3	1	2	2
36	3	1	2	2
37	3	1	2	2
38	3	1	2	2
39	3	1	2	2
310	3	1	2	2
311	3	1	2	2
312	3	1	2	2
313	3	1	2	2
314	3	1	2	2
315	3	1	2	2
316	4	1	1	1
317	RESERVED			
318	RESERVED			
319	RESERVED			
320	0	1	3	3

3. In the case that key up/down is already active (11), only in the case of adjusting the number (up/down = 2), the key affects the adjustment. So, the number increases automatically. Rate of the adjustment and the value of decrease/increase is controlled by resister ADJI.

6.3.5 *Process of Converting ΔV into Tap Number and Keys Commands*

The curve that converts value of ΔV ($\Delta V = V_{\text{ref}} - V_{\text{rms}}$) into the tap number is a step curve with hysteresis as shown in Fig. 6.19. In this curve the width of steps is 2a and the value of the curve on every interval $[(2\,k-1)\text{a}, (2\,k+1)a]$ is a constant value (k is the tap number). Limits of k variations is: $k = -24, -23, \ldots, 0, \ldots, 24$ and therefore $|k| \le k_{\max} = 24$.

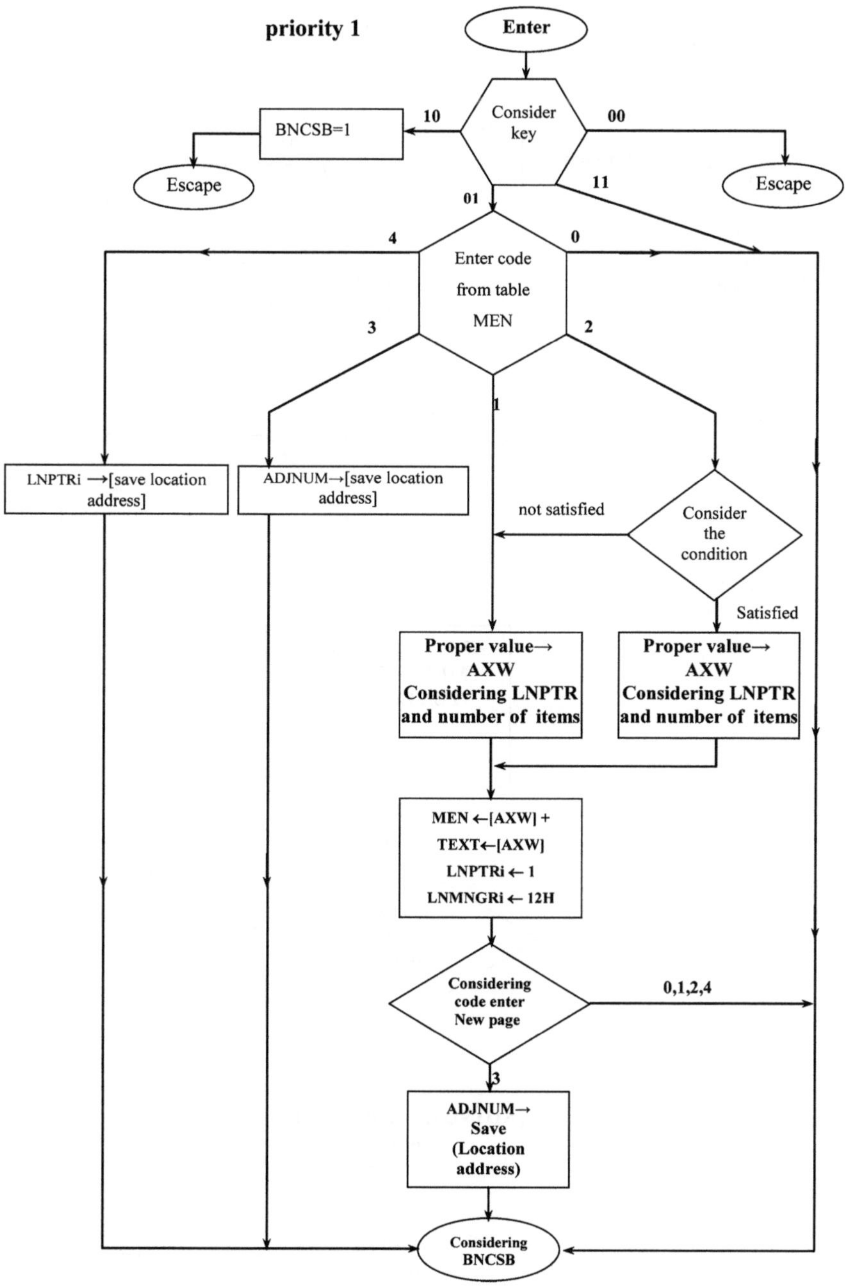

Fig. 6.15 Flow-chart of serving enter key

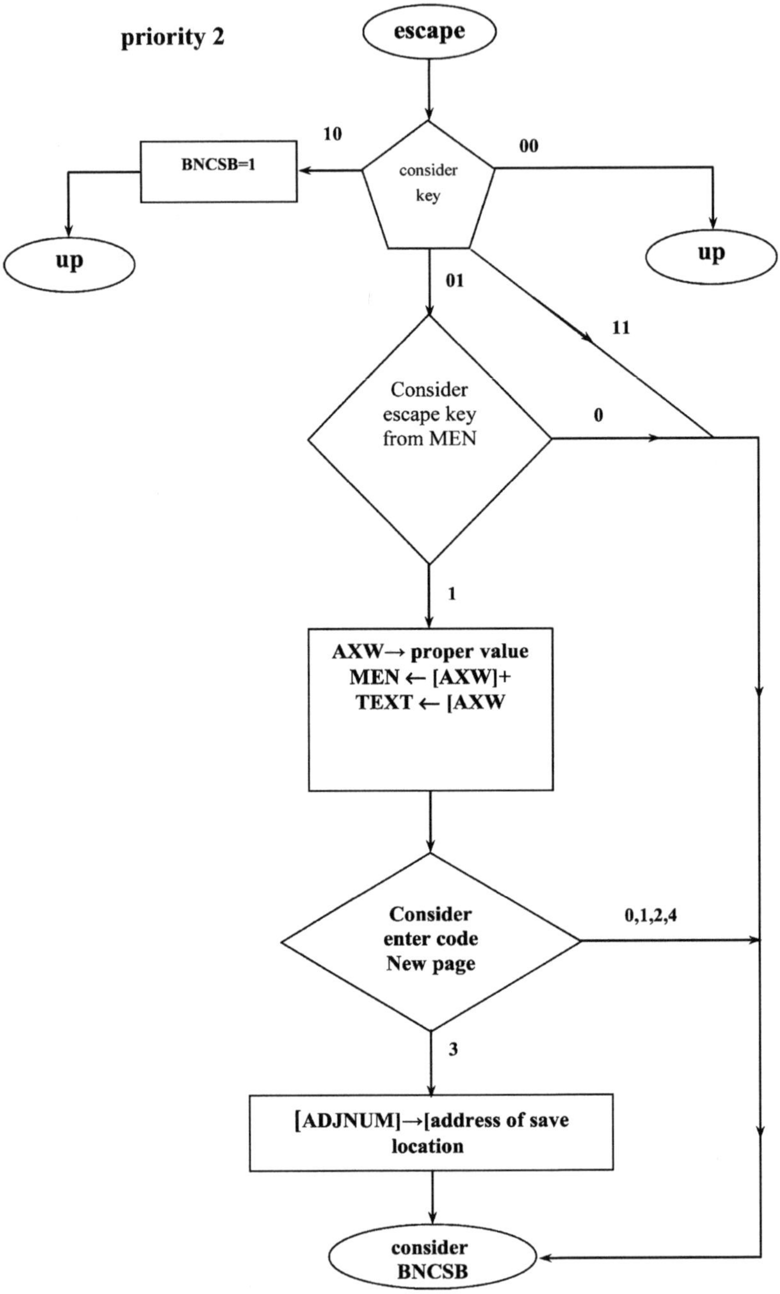

Fig. 6.16 Flow-chart of serving escape key

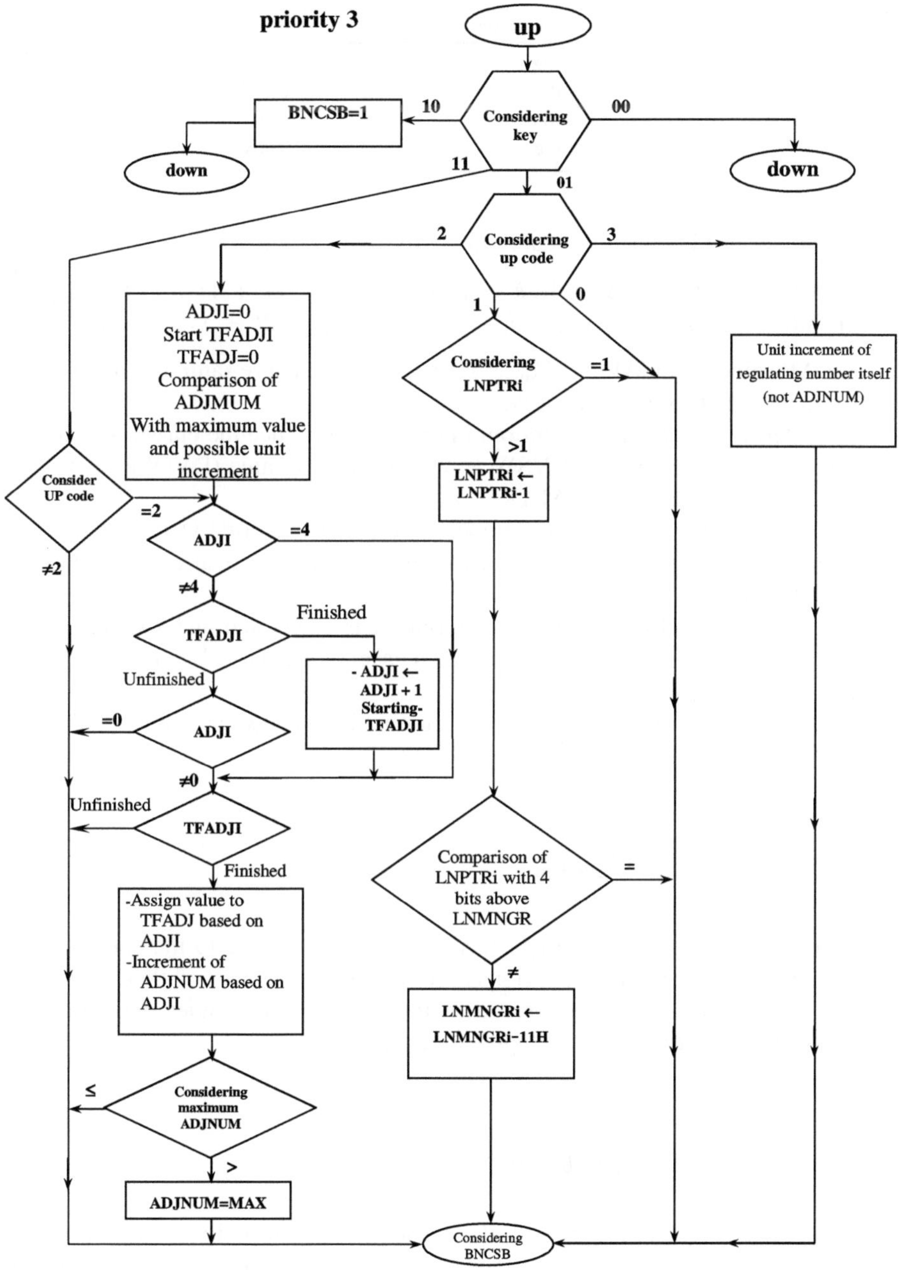

Fig. 6.17 Flow-chart of serving up key

Table 6.8 Meaning of values in study of key

Present state	Past state	
Inactive	Inactive	00
Active	Inactive	01
Inactive	Active	10
Active	Active	11

As shown in Fig. 6.19, the curve has a hysteresis with width of 2 ΔV_{offset}:

- for rising ΔV: $T = k$; $\Delta V \in [(2k - 1)\,a + \Delta V_{offset}, (2k + 1)\,a + \Delta V_{offset}[$
- for falling ΔV: $T = k$; $\Delta V \in [(2k - 1)\,a - \Delta V_{offset}, (2k + 1)\,a - \Delta V_{offset}[$

For ΔV values over $|\Delta V| > (2k_{max} + 1)a$ range, the curve saturates and its value will remain equal to its value at $|k| = 24$.

Considering the above description, in order to introduce the curve, it is enough to introduce the steps width (DVSW = 2a), hysteresis value (DVOFF = ΔV_{offset}) and maximum tap number (KMAX – k_{max}).

To avoid the use of negative numbers, the curve is shifted to the right on the horizontal axes by $(2k_{max} + 1)a$ and to the up on the vertical axes by k_{max}, so Fig. 6.20 is obtained.

Therefore, instead of ΔV in the program, x is calculated and so the value of SWPN is obtained from the curve. In such a case, the actual value of T is equal to: SWPN – k_{max}.

Process of converting ΔV into the tap number has been shown in block diagram of Fig. 6.21.

6.4 Experimental Results

A prototype electronic tap-changer has been built according to the specifications and descriptions given in the previous sections of this chapter and then it has been tested. The test results are presented in the next sections under open-loop and closed-loop performance of the tap-changer [2].

6.4.1 Open-loop Performance

The first step in testing any power electronics equipment is its open-loop performance checking. In the open-loop case, the power part of the equipment and also command signals generation part from control section are tested and their performances are investigated. To do the open-loop test, a test software has been developed and arranged as such that the tap number starts from its minimum value (−24) and then one number is added to it at any cycle (at the rising edge of the voltage digital signal), until the tap number reaches zero. At tap number 0 the

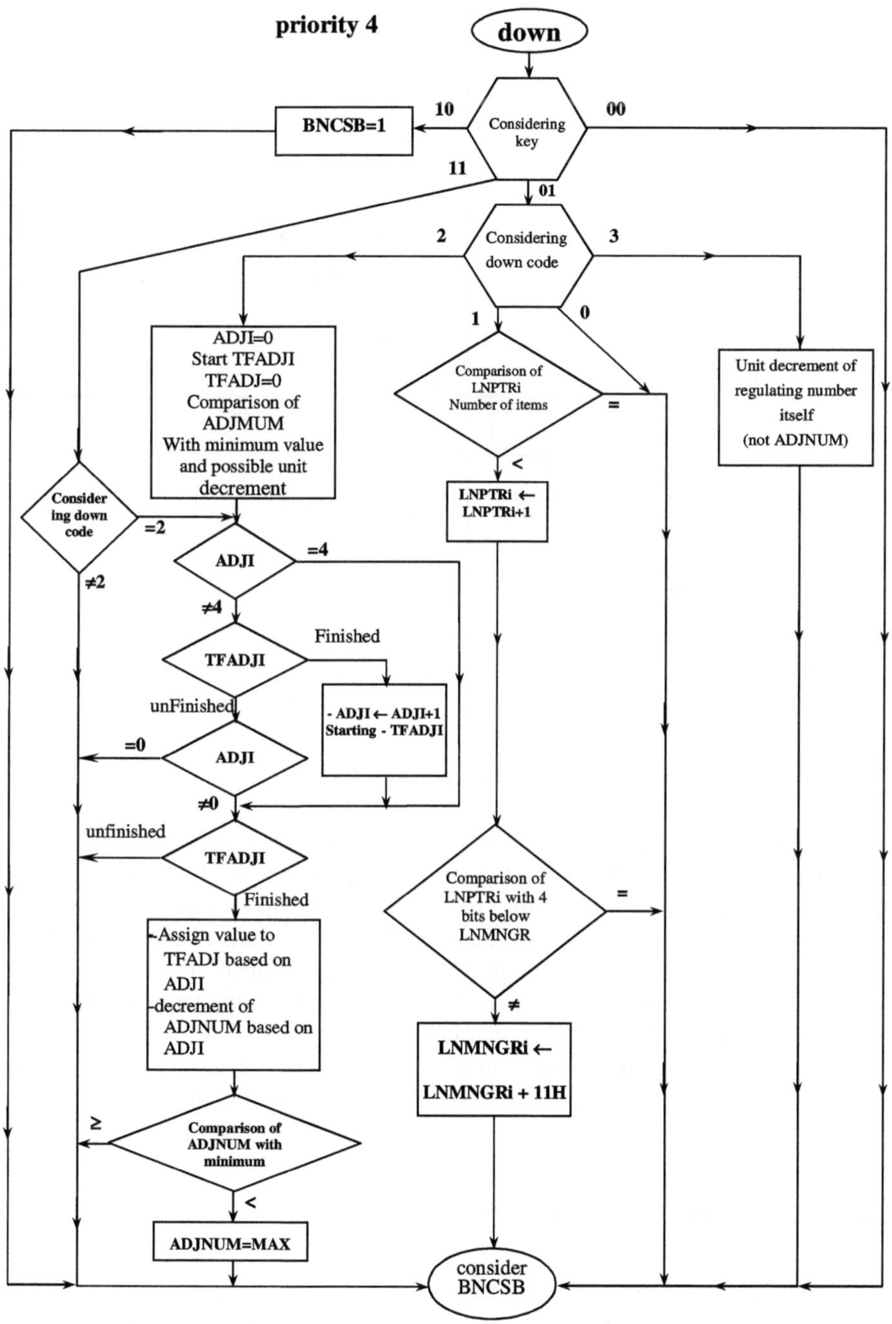

Fig. 6.18 Flow-chart of serving key down

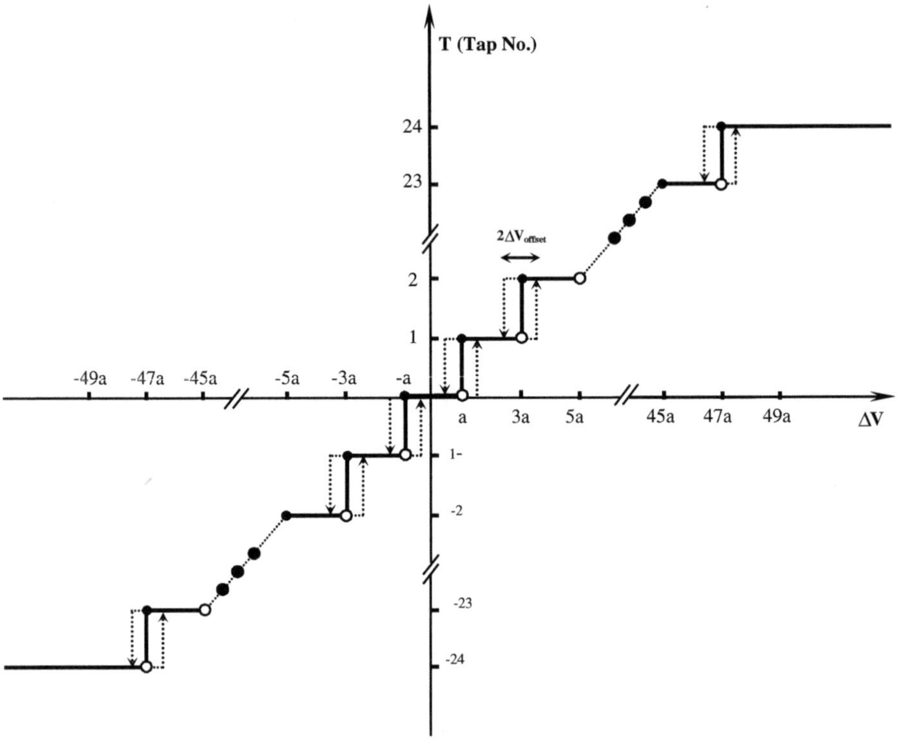

Fig. 6.19 Hysteresis curve to convert ΔV to tap number

cycle is stopped and the trend of rising the tap number continues from 0 to its maximum value (+24). When the tap number approaches +24, at the next cycle the tap number jumps to −24. Figure 6.22 presents the tap-changer output voltage in the no-load case for the above-mentioned test software. This figure shows different cases for the output voltages. Meanwhile, the linear steps of the output voltages are clear in the figure. Capability of jumping from the largest tap number to the lowest one has been also shown in Fig. 6.22.

Figure 6.23 presents the voltage output waveform for the same test software in on-load case (load current equal to 5 A). The voltage drop is about 1 V.

6.4.2 Closed-loop Performance

Following the open-loop test and verifying the power part and, switches drivers and also command signals generation part, the closed-loop test is carried out in order to show the capabilities of the closed-loop control system in output regulation. In the following parts, the closed-loop performance is presented under the line regulation and load regulation,

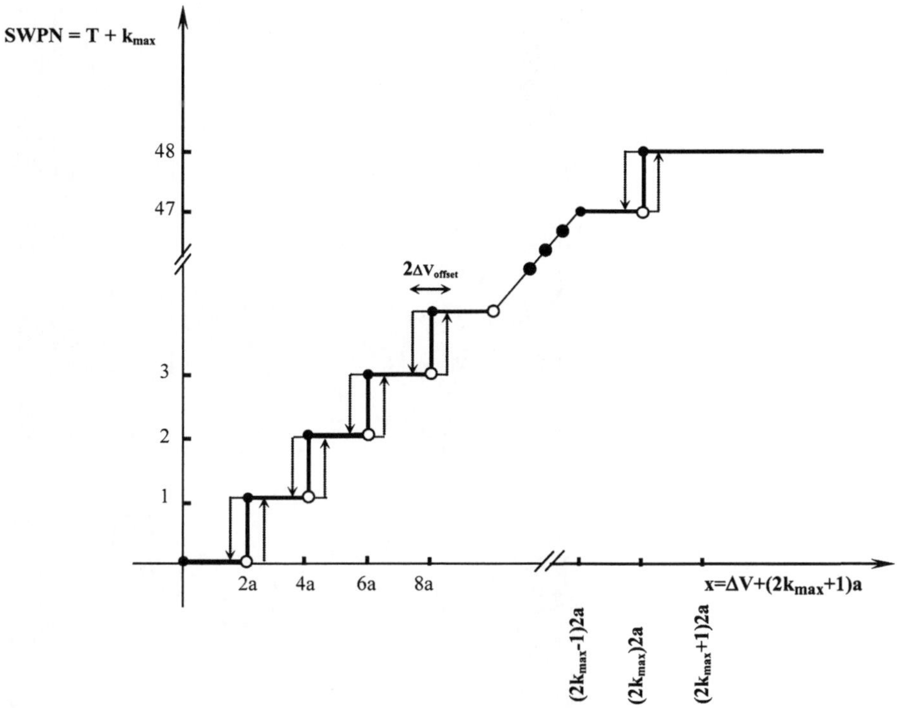

Fig. 6.20 Shifting curve of Fig. 6.19 to avoid negative values

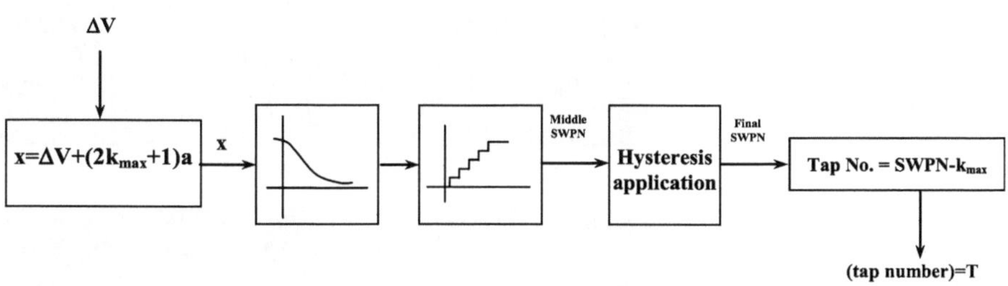

Fig. 6.21 Block diagram of process of converting ΔV to tap number

6.4.2.1 Line Regulation

The purpose of the line regulation is adjusting the output voltage against input voltage changes. The case of the examination of the line regulation, the load current is fixed. Figures 6.24, 6.25, and 6.26 present the tap-changer output voltages for input voltages equal to 176 V, 206 V and 241 V respectively. The tap numbers in these figures are equal to 23, 4 and −12 respectively. These three figures show that in spite of changing the input voltage, amplitudes of the output voltages have been remained constant. Figures 6.27, 6.28, and 6.29 show the

Fig. 6.22 Tap-changer output voltage waveform in no-load case using test software (open-loop)

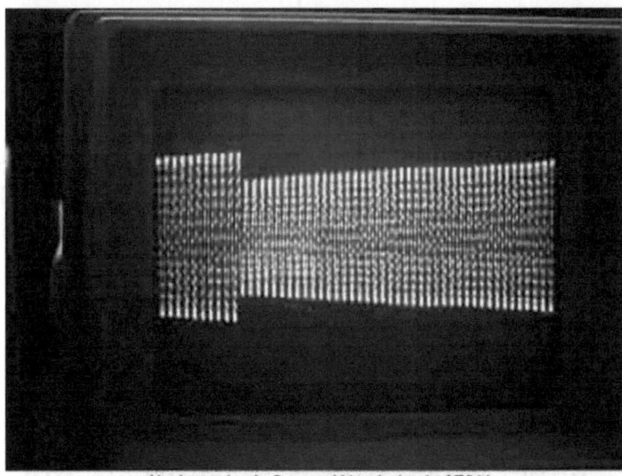

Fig. 6.23 Tap-changer output voltage waveform for 5 A load case using test software (open-loop)

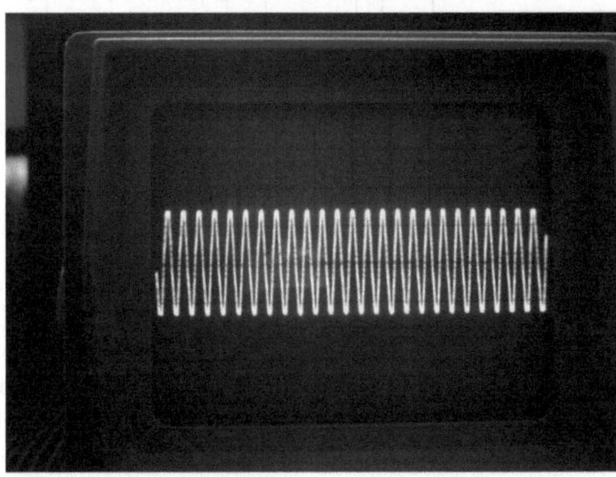

Fig. 6.24 Tap-changer output voltage waveform for input voltage 176 V in no-load case

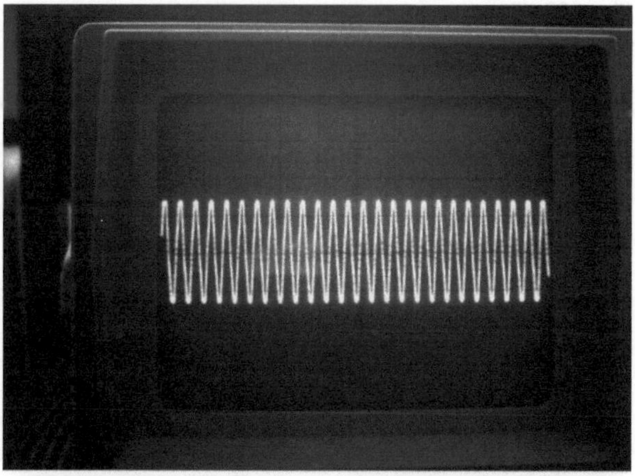

Fig. 6.25 Tap-changer output voltage waveform for input voltage 206 V in no-load case

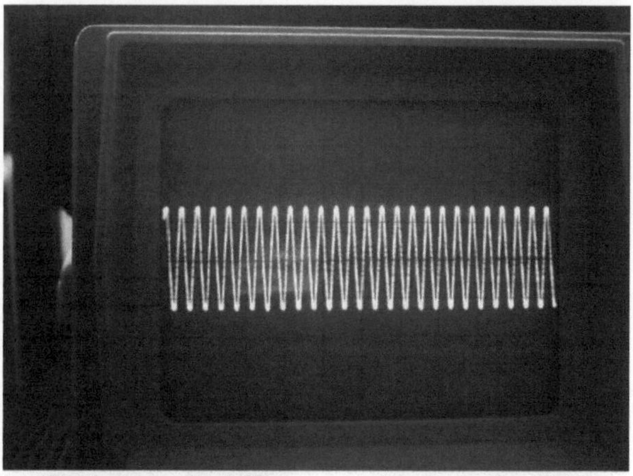

Fig. 6.26 Tap-changer output voltage waveform for input voltage 241 V in no-load case

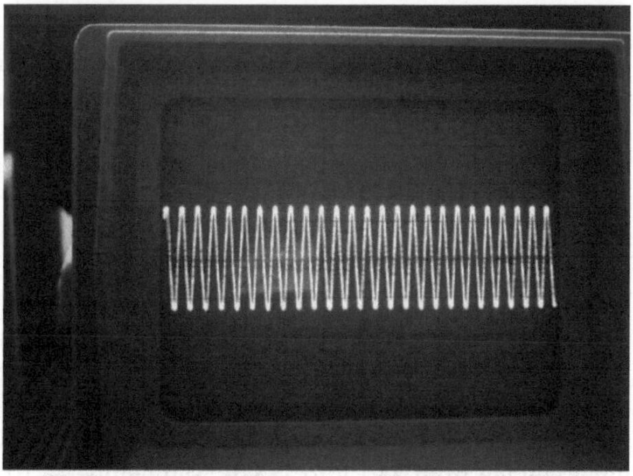

Fig. 6.27 Tap-changer output voltage waveform for input voltage 179 V and 5 A load case

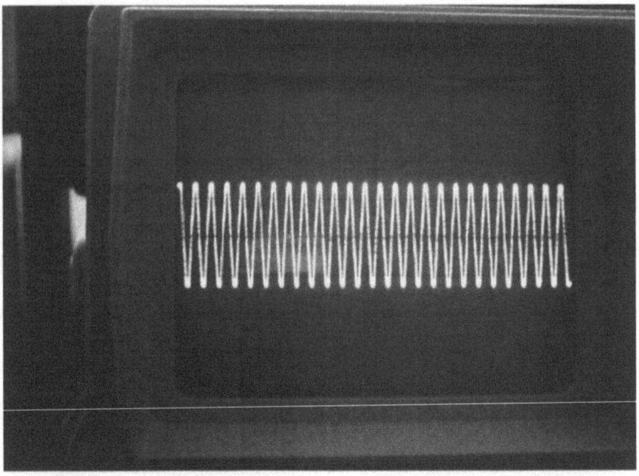

Horizontal axis: 0.05 s Vertical axis: 220 V

Fig. 6.28 Tap-changer output voltage waveform for input voltage 206 V and 5 A load case

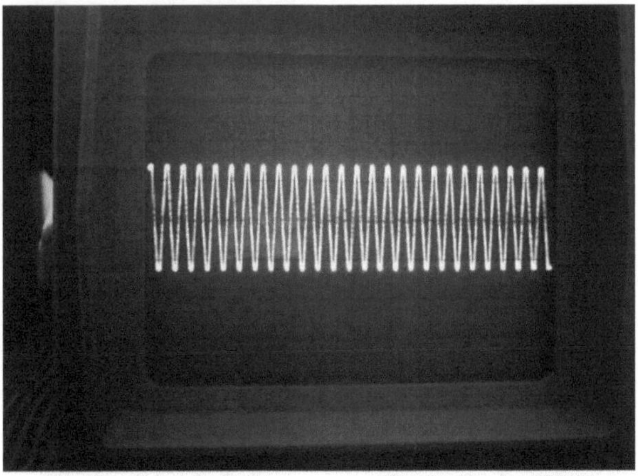

Horizontal axis: 0.05 s Vertical axis: 220 V

Fig. 6.29 Tap-changer output voltage waveform for input voltage 241 V and 5 A load case

tap-changer output voltages at the case of 5 A load current for input voltages equal to 179 V, 206 V and 241 V respectively. The tap numbers in these figures are equal to 23, 6 and −11 respectively. It is clear from these figures that the output voltages are fixed. Figure 6.30 shows the primary current waveform in this case.

Up to now, the line static regulation has been shown. Now input voltage is manually changed by an auto-transformer and transient output voltage is examined. This is called the line dynamic regulation. In all following tests the integrator factor has been fixed on 10, except the opposite is mentioned. Figure 6.31 presents the output voltage waveform in the no-load case for quick rising of the input voltage from 175 V to 240 V. Tap number 23 changes to −10. It is observed that the output voltage returns to its initial value after about 5 cycles. Figure 6.32 shows the output voltage waveform in the no-load case for a quick dropping of the input voltage from 240 *V* to 188 *V*. Tap number −10 changes to 16. In this case it

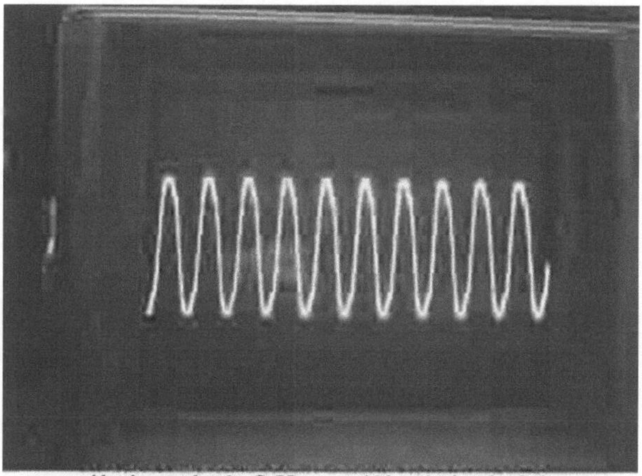

Fig. 6.30 Transformer primary current waveform for 5 A load case

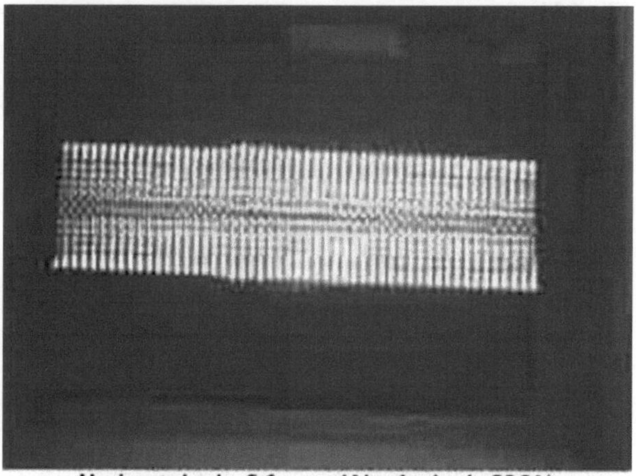

Fig. 6.31 Tap-changer output voltage waveform in no-load case for quick rise of input voltage from 175 V to 240 V

is clear that the output voltage returns its initial case after about 5 cycles. Waveforms of Figs. 6.33 and 6.34 present the output voltage waveforms at load current of 5 A for input voltages of 178 *V* to 239 *V* and from 250 V to 195 *V* respectively. The tap numbers in the above-mentioned cases are changed from 23 to −10 and −14 to 12 respectively.

In order to present the large changes of the tap due to the transient mode of the power supply and noises of the system, a low-pass filter software has been placed after comparison of the rms value with the reference value. The output voltage waveforms for quick changes of the input voltage including this filter have been shown in Figs. 6.35, 6.36, 6.37, 6.38). It is clear that the closed-loop system will be second-order by adding this filter and a few oscillations in the responses are visible.

Figure 6.39 shows the tap-changer output voltage waveform for integrator factor equal to 50. As was predicted, in this case the output voltage is oscillatory.

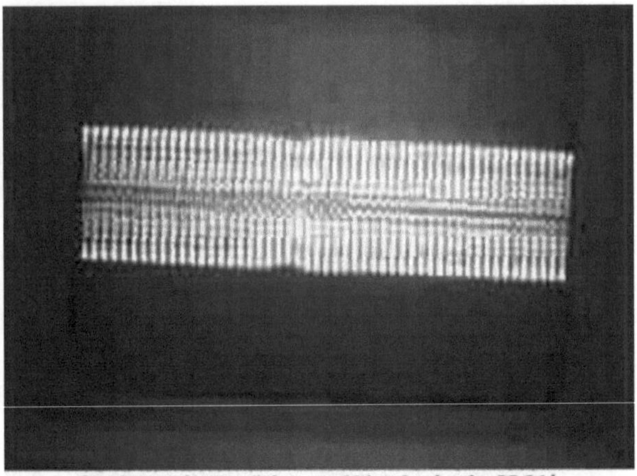

Fig. 6.32 Tap-changer output voltage waveform in no-load case for quick drop of input voltage from 240 V to 188 V

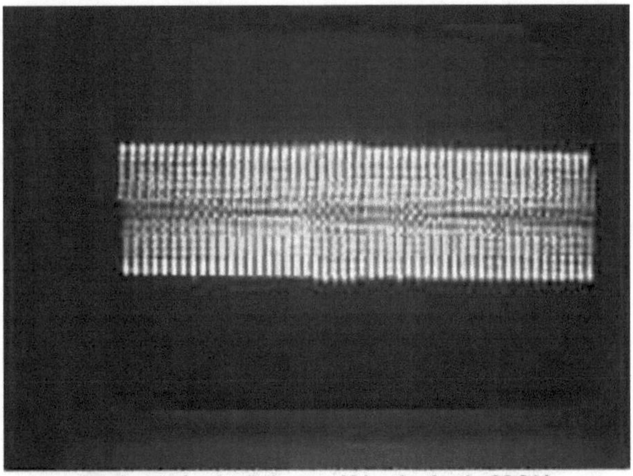

Fig. 6.33 Tap-changer output voltage waveform for 5 A load and quick rise of input voltage from 178 V to 239 V

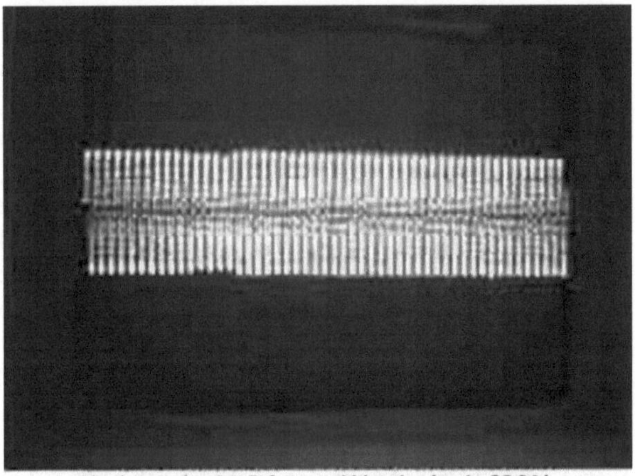

Fig. 6.34 Tap-changer output voltage waveform for 5 A load and quick drop of input voltage from 250 V to 195 V

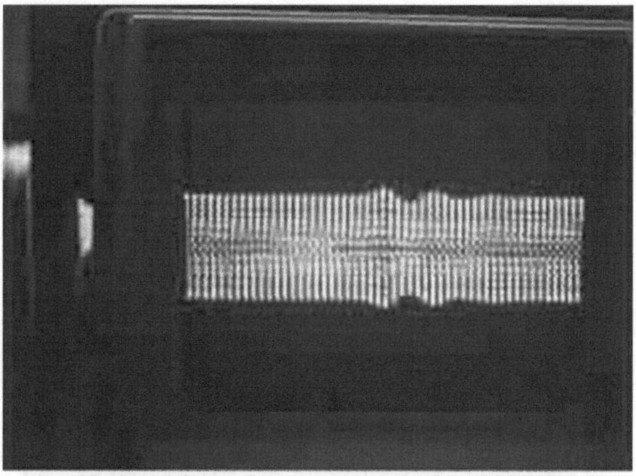

Fig. 6.35 Tap-changer output voltage waveform in no-load case and for quick rise of input voltage from 170 V to 220 V with low-pass filter software

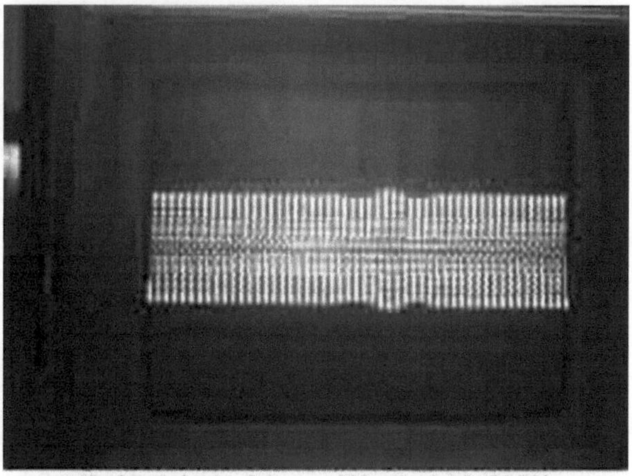

Fig. 6.36 Tap-changer output voltage waveform in no-load case and for quick drop of input voltage from 240 V to 198 V with low-pass filter software

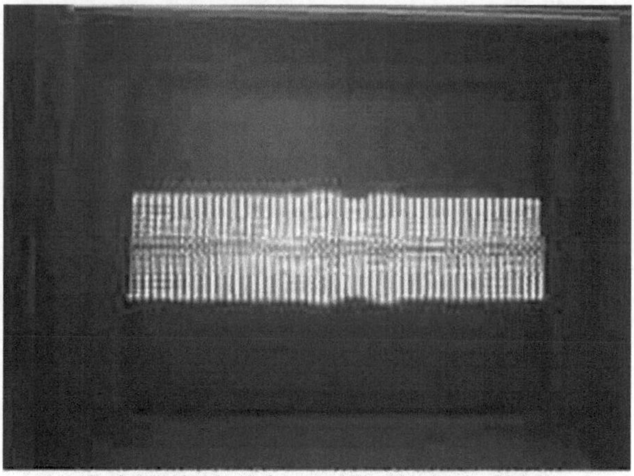

Fig. 6.37 Tap-changer output voltage waveform for 5 A load and quick rise of input voltage from 181 V to 233 V with low-pass filter software

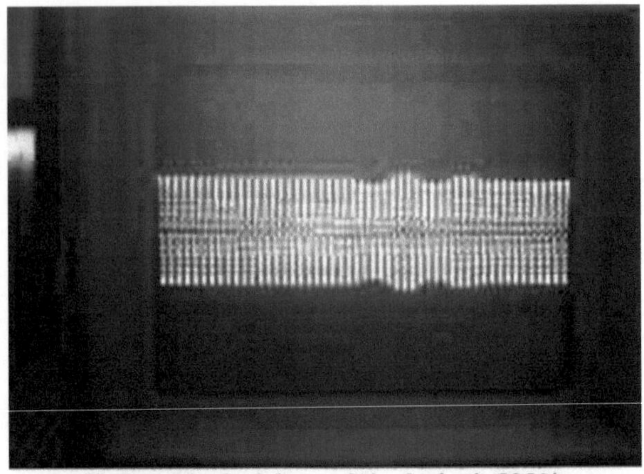

Horizontal axis: 0.1 s and Vertical axis 220 V

Fig. 6.38 Tap-changer output voltage waveform for 5 A load and quick drop of input voltage from 238 V to 199 V with low-pass filter software

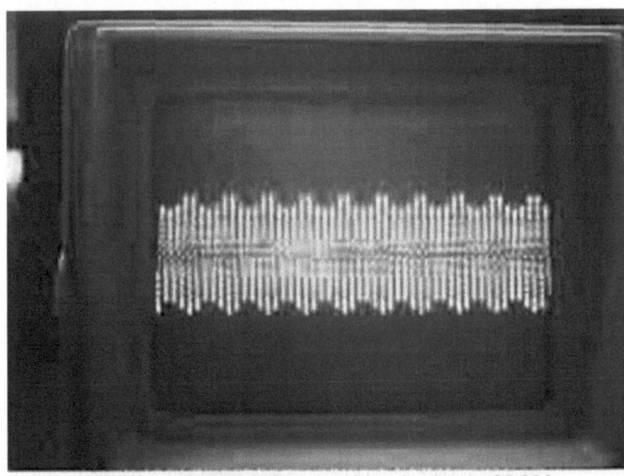

Horizontal axis: 0.1 s and Vertical axis 220 V

Fig. 6.39 Tap-changer output voltage waveform for integrator factor of 50

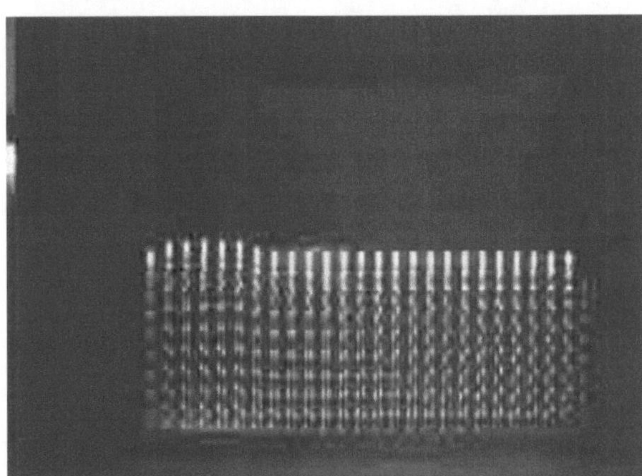

Horizontal axis: 0.1 s and Vertical axis 220 V

Fig. 6.40 Tap-changer output voltage for change of load current from 5 A to 0 at input voltage 224 V

Fig. 6.41 Tap-changer output voltage for change of load current from 0 to 5 A at input voltage 224 V

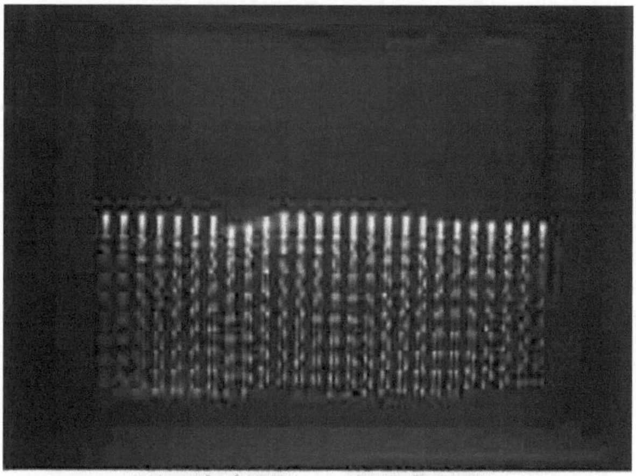

6.4.2.2 Load Regulation

The load regulation is fixing the output voltage against the load current changes. In this case, the input voltage is kept constant. Figures 6.40 and 6.41 present the output voltage changes for sudden interruption and sudden transformer energizing at input voltage equal to 224 V respectively. In these waveforms, the output voltages return to their initial value after 5 cycles. It is noted that in order to show the changes of the output voltages in these two figures, the vertical axes scale are taken to be larger.

References

1. Faiz J, Siahkolah B (2008) Implementation of a low power electronic tap-changer in transformers. IET J Electric Power Appl 2(6):362–373
2. Faiz J, Siahkolah B (2011) Solid-state tap-changer of transformers: design. Control and implementation. Int J Electrical Power Energy Syst 33:210–218

Chapter 7
Conclusions and Suggestions

7.1 Conclusion

The most important results described in the previous chapters are briefly expressed in this chapter.

- Application of the full-electronic tap-changer is preferred to the conventional mechanical tap-changer. The reasons are discussed below.

  1. Low Cost of Maintenance and Service

  Tap changing in the mechanical tap-changer generates arc in the contacts of the diverter switches. This arc causes contamination of the oil surrounding the diverter switches and also leads to the erosion of their contacts. Meanwhile, the whole tap changing process in the mechanical tap-changer is basically performed mechanically. Therefore, in the mechanical tap-changers the conditions of oil, contacts and mechanical moveable parts must be examined regularly and are serviced if necessary. However, there is no arc in the full-electronic tap-changer during the tap-changing and also no moveable mechanical part. So, their maintenance cost is very low and almost zero.

  2. High Speed

  The tap-changing process in the mechanical tap-changers is slow because of its mechanical nature and also the required time for storing the necessary energy for an appropriate implementation of the tap changing process. In the resistor tap-changers, tap changing from tap number 1 to tap number 19 takes about 100 s, while this takes μs in the full-electronic tap-changers. The reason is that the switching in solid-state switches is very high, which leads to the high speed of the tap changing process.

J. Faiz and B. Siahkolah, *Electronic Tap-changer for Distribution Transformers*, Power Systems, DOI: 10.1007/978-3-642-19911-0_7,

3. Jumping in Tap-changing

In the full-electronic tap-changers, the circulating current between the taps is basically zero. Therefore, jumping in the tap changing is possible. There is not such a choice in the mechanical tap-changers due to their mechanical limits and circulating current.

4. Better Capability and Performance

High speed and controllability of the solid-state tap-changers and also lack of mechanical limits in the switches arrangement lead to better capability and performance of the full-electronic tap-changers. Some of these capabilities are as follows:

(a) Achieving higher number of steps, tap number and also lower solid-state power switches; these capabilities arise from the lack of limitation in the solid-state power switches arrangement.
(b) Proposing the full-electronic tap-changer as a fast static regulator such that it is considered as a custom power tools in power quality. It can mitigate sag and also flicker in power systems.

In order to replace the mechanical tap-changers by the full-electronic tap-changers, the latter tap-changers must have the following features:

1. Low Cost

Application of the solid-state power switches in the full-electronic tap-changers raises their cost. So, to modify the price, it is necessary to use the low number of switches and minimum voltage and current.

2. High Reliability
3. Standing higher than the rated current and voltage in the fault conditions in network

For the fault conditions of the network (such as short circuit or voltage rise due to the lightning etc.). The voltage and current become higher than their rated values. The full electronic tap-changer must be designed such that it could stand these conditions and does not burn.

- Since reducing the costs of the full-electronic tap-changer is the most important aim in its design, the main goals in the design of the power part of the tap-changer are taken as follows:
  1. Minimal number of the solid-state power switches and their voltages and currents.
  2. Minimal number of the transformer taps.
  3. Maximal Number of the voltage steps in the output voltage regulation range in order to increase the regulation accuracy.

To satisfy the above-mentioned goals, the process of designing the power section of a full-electronic tap-changer is divided into two steps:

Step 1: Design of the winding taps structure as such that criterion C_1 is maximized:

$$C_1 = \frac{\text{Achievable voltage steps number}}{\text{Transformer taps number}} \tag{7.1}$$

Step 2: Switches configuration design for the designed winding taps in the first step is as such that criterion C_2 is maximized:

$$C_2 = \frac{\text{Realizable voltage number}}{\text{Number of switches}} \tag{7.2}$$

- According to criterion C_1, the optimal structure of the taps winding is as shown in Fig. 7.1 and its related criterion is $v^i/3i$.

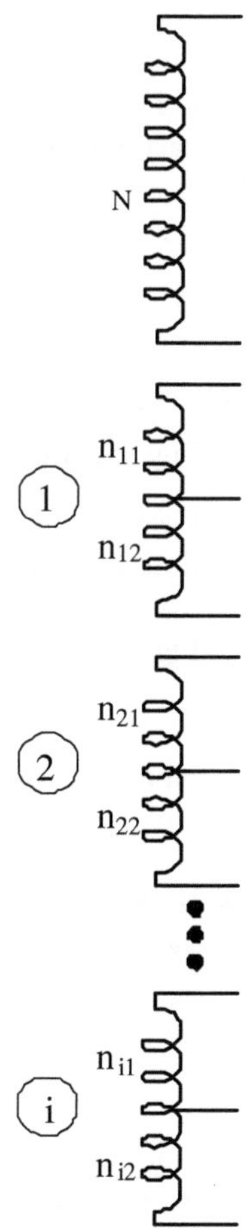

Fig. 7.1 Optimal structure for taps winding for criterion C_1

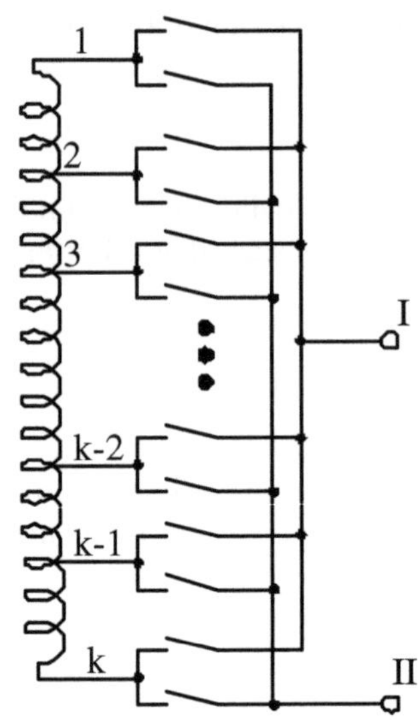

Fig. 7.2 Optimal configuration for switches based on criterion C_2

- For non-repetitive achievable voltages in optimal structure of Fig. 7.1 which vary linearly with equal steps, the following steps must be satisfied:

$$V_{n_{j1}} = 7^{j-1} V_{\text{step}}; \quad V_{n_{j2}} = 2 \times 7^{j-1} V_{\text{step}}; \quad \text{j} = 1, 2, \ldots, i \tag{7.3}$$

  where V_{nj1} and V_{nj2} are the voltages corresponding with windings $nj1$, $nj2$ and V_{step} is the achievable step voltages.

- According to criterion C_2, the optimal configuration of switches is as shown in Fig. 7.2 and the related criterion is equal to $\frac{K^2-K+1}{2K}$ where K is the taps number.

- Considering the optimal structure of taps winding (Fig. 7.1) and optimal configuration of Fig. 7.2, the optimal topology of the power part of the full-electronic tap-changer is suggested as shown in Fig. 7.3. In the optimal topology of Fig. 7.3, the governing equations are:

$$\begin{aligned} \text{Number of taps}(N_T) &= 3i \\ \text{Number of solid-state bidirectional switches } (N_S) &= 6i \\ 2V_{nj1} = V_{nj2} &= 2 \times 7^{j-1} V_{\text{step}} \end{aligned} \tag{7.4}$$

- If the cost criterion for the power part of the full-electronic tap-changer is expressed as follows:

$$C_3 = \alpha_{N_T} N_T + \alpha_{N_S} N_S \tag{7.5}$$

  where N_T is the taps number, N_S is the solid-state power switches (bi-directional) number and αN_v and αN_S are the related weighted cost of each tap and switch respectively. So, the topology of Fig. 7.3 will have the minimal C_3 where

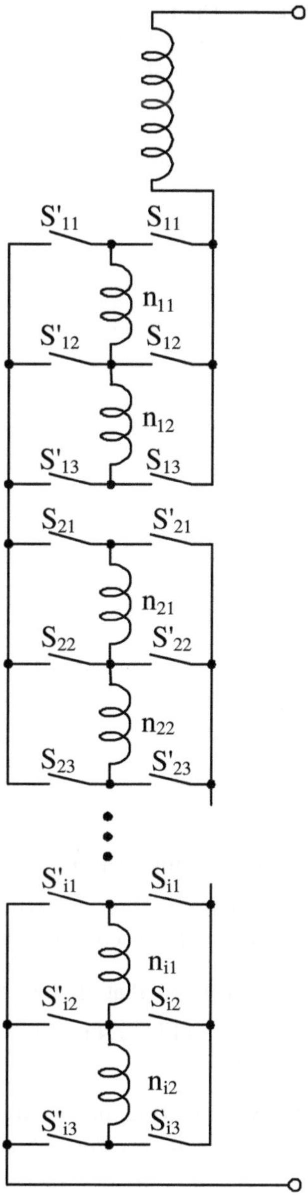

Fig. 7.3 Suggested optimal topology for power part of a full-electronic tap-changer-combination of optimal structure of windings of Fig. 7.1 and optimal configuration of switches of Fig. 7.2

$\frac{\alpha_{N_S}}{\alpha_{N_T}} < 3.37$. It means that the price of adding a bi-directional solid-state switch is less than 3.37 times of the adding a tap. If this condition does not satisfy, then criterion C_3 related to Fig. 7.4 will be minimal.

- If in the estimation of the price of the power section the full-electronic tap-changer, the voltage of the solid-state switches also intervene, criterion C_4 can be suggested for cost estimation:

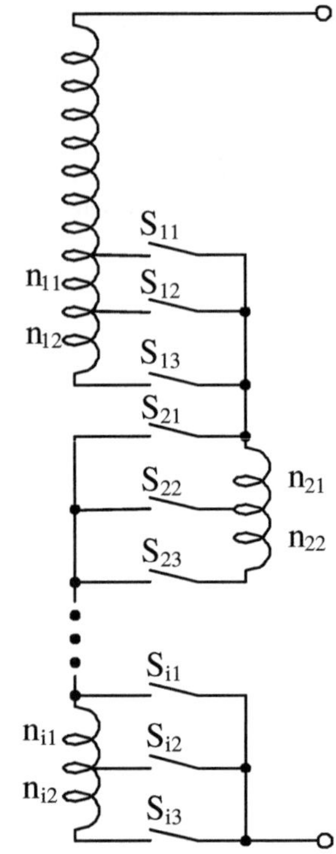

Fig. 7.4 Optimal topology according to cost criterion C_3 with condition of: $\frac{\alpha_{N_S}}{\alpha_{N_T}} > 3.37$

$$C_4 = \alpha_{N_T} N_T + \alpha_V \sum_{j=1}^{N_S} V_j \qquad (7.6)$$

- Considering criterion C_1, if $\frac{\alpha_N V_{step}}{\alpha_{N_T}} < X$, topology of Fig. 7.3 and if $\frac{\alpha_N V_{step}}{\alpha_{N_T}} > X$, topology of Fig. 7.4 will be optimal. Value of X depends on the value of N_V. Table 7.1 presents the values of X for three values of N_V.

It is noted that if the condition $\frac{\alpha_N V_{step}}{\alpha_{N_T}} > X$ is satisfied, in the best conditions, topology of Fig. 7.4 is only 6% better than that of Fig. 7.3, from cost criterion C_1 point of view.

- By comparing the possible suggestions for realization of the bi-directional solid-state switches, Fig. 7.5 has been selected:

Table 7.1 Values of X for different N_V

N_V	X
25	1
50	0.6
75	0.5

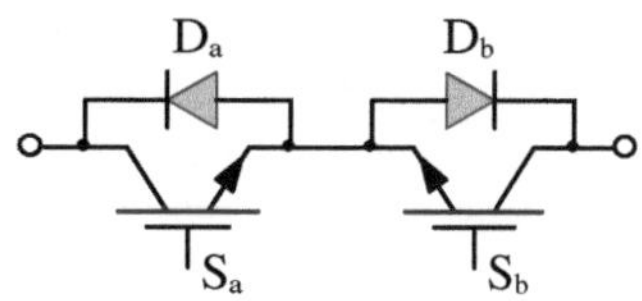

Fig. 7.5 Selected structure for realization of bi-directional solid-state switches

It is also possible to use the fast solid-state switches such as MOSFET and IGBT in Fig. 7.5, because there is the capability of controlling the direction of the current.

- Assuming a prefect sinusoidal current, efficiency of the electronic tap-changer is obtained as follows:

$$\eta = 1 - 0.9 \times \frac{(V_\gamma + V_{\mathrm{CE_{sat}}})}{V_n} \times n \tag{7.7}$$

where n is the number of the bi-directional switches in the current path, V_n is the rated voltage of the transformer in the tap-changer side. Meanwhile, V_γ is the voltage drop on diodes D_a or D_b in the on state and $V_{CE\mathrm{sat}}$ is the drop of switch-in states of switches S_a or S_b. For example if the following values are considered for the above-mentioned parameters, then efficiency of the electronic tap-changer will be equal to 99.89%.

$$n = 6,\ V_\gamma + V_{\mathrm{CEsat}} = 4\,\mathrm{V},\ V_n = 20\,\mathrm{kV} \tag{7.8}$$

- Considering the following reasons, the current commutation process in the bi-directional switches must be carried out with specific care:
  1. During the commutation, both bi-directional switches must not be switched-off simultaneously. This means closing the current path and creating high voltages.
  2. During the commutation, both di-directional switches must not be switched-on simultaneously. This means the short circuiting the taps and creating the high circulating currents.

Therefore, during the dead-time commutation, the over-lapping time are not permissible, so, the usual commutation methods in bi-directional switches cannot be used.

- By comparing two commutation methods in the bi-directional switches, the commutation method based on the voltage values has been selected to use here. Table 7.2 compares these two commutation methods briefly according to the current direction and voltage value.
- Considering leakage inductance of taps windings L_{lt}, there will be an over voltage equal to $L_{\mathrm{lt}}\alpha$ on the solid-state switches where α is the slope of current dropping to zero in the solid-state switches, following their zero gate command. To damp this over voltage, inserting a capacitor C on the tap terminals is

Table 7.2 Comparison of commutation methods according to current direction and voltage value in bi-directional switches

No.	Commutation type	Current sensor	Sensitive to current direction change during commutation	Short circuit risk due to measurement error	Required number of stages in commutation
1	Current direction base	Yes	Yes	Yes	4 stages can be reduced to 2 stages
2	Voltage value base	No	No	Yes	4 stages

suggested. By inserting this capacitor, the over voltage will be equal to $I_L\sqrt{\frac{L_{lt}}{C}}$ where I_L is the load current at the commutation time. For example, if $L_{lt} = 200\ \mu H$ and $I_L = 40$ A, then $C = 8\ \mu F$ can decrease the over voltage to 200 V, while for non-existence of this capacitor this over voltage will be equal to 20 kV assuming $\alpha = 100$ A/μs and this certainly can damage the solid-state switches.

- If commutation occurs only at the zero-crossing load current, then the over voltage at the commutation will be zero (even without snubber capacitor), but the switching losses will also move towards zero. So, the tap must be changed only at the load current zero-crossing instants.
- For comparing the available methods in the modulation of switches of full-electronic tap-changer, no-commutation case has been selected here. Table 7.3 summarizes the comparison of the modulation methods.
- By comparing the performance characteristics of the mechanical and full-electronic tap-changers it is obvious that the following differences are essential for the aims and duties of the controller of these two tap-changers.
    1. Decreasing frequency of tap-changes in the controller of the mechanical tap-changers is considered as the design goals, while this is not so in the full-electronic tap-changers.

Table 7.3 Comparison of different modulation methods in a full-electronic tap-changer

No.	Modulation method	Output additional harmonics	Voltage regulation and control	Possible voltage waveform modification	Switching losses
1	Phase control	Low-frequency	Continuous	No	Average
2	Discrete pulse	Low frequencies	Non-continuous	No	Average
3	Pulse width	High frequencies	Continuous	Yes	High
4	No modulation	No additional harmonics	Non-continuous	No	Low

2. In the controller of the mechanical tap-changers, any tap position can be changed to the previous or next taps (step by step process), while there is no such limitation in the full-electronic tap-changers.
3. In the controller of the full-electronic tap-changers, it is essential to have a lookup table to store the position of any of the switches corresponding to each output states due to the complicated configuration of taps and also tap-changer switches, while there is no need to have such memory in the mechanical tap-changers.
4. The tap-changing rate is high in the full-electronic tap-changers, therefore it is necessary to apply an algorithm for detecting the amplitude and phase of the sinusoidal variables (such as primary voltage, secondary voltage and load current) quickly and precisely, while tap-changing process in the mechanical tap-changers are slow and averaging methods over a few cycles is suffice.
5. To realize the soft switching conditions in the controller of the full-electronic tap-changers, it is essential to detect the zero current crossing instants and let the tap changing in these instants; however, in the mechanical tap-changers the tap- changing process is slow and such procedure cannot be applied.

- In order to prevent the oscillation of the control system of the full-electronic tap-changer, a dead-zone block is set following the comparison of the amplitude of the output voltage and reference voltage. Dead-band (DB) in this block must satisfy the following condition:

$$2DB \geq \frac{V_{1\max}}{a_\circ} \times \frac{\frac{\Delta a}{a_\circ}}{1 + (-2n_{\max} + 1)\frac{\Delta a}{a_\circ}} \tag{7.9}$$

where $n_{\max}$ is the maximum of the tap number, a_0 is the transformer turn ratios, Δa is the step variations of a with one tap change and $V_{1\max}$ is the maximum of the primary voltage. Meanwhile, it is noted that increasing DB more than the limit given by Eq. 7.7 causes a permanent steady-state error in the output which is undesirable.

- To design the compensator in the control system of the full-electronic tap-changer, the following new method was suggested:

First the design problem is fully simplified by the following simplifying assumptions; then this simplified problem is solved. Finally, these assumptions are concealed one by one and the required modification is applied to the answer in order to achieve the solution of the main problem.

1st Assumption: All quantities of the system are always sinusoidal.
2nd Assumption: All impedances of the system are ohmic.
3rd Assumption: Tap changing time is almost zero and the taps can be changed any time.
4th Assumption: Variations of a is continuous.
5th Assumption: There is no delay in the estimation of amplitude and phase of the sinusoidal quantities of system.

- The simplified problem can be solved by an integral compensator. This integral compensator compensates the effect of V_1 changes and $(Z_T + Z_C)_2I$ (series impedances drop). The integrator gain determines the system band-width and therefore its speed. Meanwhile, it was noted that the use of feed-forward path for quick selection of an appropriate value for *a*.

- The designed controller will still be valid by concealing the 2nd assumption. In this case condition 7.8 is essential for the system stability:

$$\left|\frac{V_1}{a}\right| = |A(V_{\text{ref}} - |V_{2m}|)V_1| > |ZI_2\cos(\theta_z + \theta_I)| \tag{7.10}$$

  Z is the phasor of $\widehat{Z}_T + \widehat{Z}_C$ and θ_Z is its phase. Meanwhile, I_2 is the phasor of the load current, θ_I is its phase and *A* is the compensator gain.

- By concealing the 4th assumption, it was clear that after estimation of ΔV (difference between the reference and output voltages), it is necessary to add the dead-zone block in the control system. The *DB* in this block is evaluated using Eq. 7.7.

- Assuming the use of a quick detection method for the amplitude and phase of the sinusoidal quantities as such that it could extract the fundamental component from other harmonics, it is also possible to conceal the 1st assumption without any damage to the design.

- Finally, if the 3rd assumption is concealed, it is essential to take a specific care about the selection of the integrator gain. Otherwise, the system will become oscillatory.

- From the comparison of the proposed methods for estimation of the amplitude and phase of the sinusoidal quantities, a numerical matrix method was chosen here. Table 7.4 summarizes the above-mentioned methods.

- Full-electronic tap-changer system with its control part was simulated using Simulink software and simulation results confirmed the accuracy of the designs in the control part of the tap-changer. In this simulation, the behavior of the system against sudden changes in the input voltage and also load current were studied and competence and quickness of system for compensating these changes and quick approaching the output reference were confirmed.

- Considering the capability of the full-electronic tap-changer, this equipment was proposed in power quality as one of the custom power tools (series type) and particularly one of used equipment for mitigation of the voltage sag (which is the most important in power quality problems).

- If all impedances of the system are assumed resistive and there is no limit for *a*, the full-electronic tap-changer compensation is obtained approximately as follows:

Table 7.4 Comparison of different methods for computation of amplitude and phase of sinusoidal quantities

Number	Evaluation criterion-routine name	Maximum measurement delay	Computation capability amplitude	Phase	Sensitivity to additional harmonics	Complexity of method	Notes
1	Use of sinusoidal variable peak	An half cycle	Yes	No	Yes	Very simple	
2	Computation of rms value of sinusoidal variable	An half cycle	Yes	No	Yes	Simple	Rms value is calculated over an half cycle
3	90° delay of sinusoidal variable	A quarter cycle	Yes	No	Yes	Simple	
4	Fourier transform of sinusoidal variable	One cycle	Yes	Yes	No	Complicated	Sampling over a period
5	Use of PLL	t_s	No	Yes	Yes	Complicated	t_s: time for locking PLL
6	Numerical matrix method	$2NT_s$	Yes	Yes	No	Medium	N: harmonics no. Ts: sampling period

$$V_{\text{sag}} \geq 2\sqrt{\frac{Z_{\text{eq}}}{Z_L}} V_{2n} \tag{7.11}$$

where V_{2n} is the secondary rated voltage, Z_{eq} is the network equivalent impedance during the primary fault, Z_L is the load impedance and V_{sag} is the network equivalent voltage during the primary fault.

As shown in Eq. 7.11, a lower $\frac{Z_{\text{eq}}}{Z_L}$ ratio will extend the compensation range of the full electronic tap-changer, as such that if the series impedance of the network is kept zero, the tap-changer will compensate V_{sag} close to zero.

- Computations for evaluation of the compensation range of the full electronic tap-changer in real conditions (non-zero impedance and limitation for a) lead to the following results:

  1. Compensation range is changed by varying the load impedance angle. This compensation is minimal when the angle is equal to the summation of the series impedances.
  2. Range of compensation rises for capacitive loads.
  3. Range of compensation decreases by increasing the load current.
  4. Range of compensation increases by increasing the input voltage.

- An economical method was introduced to compare the full electronic tap-changer with other sag mitigation tools and this method was shown by an example. The evaluation results confirmed that the full-electronic tap-changers cost, their high efficiency and very low maintenance costs provide the most economical equipment for the sag mitigation if distribution density of various sags in a particular case are such that the amplitude of the most sags is higher than 50%.
- To verify the whole design stage (power and control parts) and also simulation results, a low-power (5 kW) prototype full-electronic tap-changer was built and tested. Experimental results confirmed the designs and theoretical predictions based on the simulations.

7.2 Suggestions

Further studies are possible in the full-electronic tap-changer which are as follows:

- Designing essential protections for the full-electronic tap-changer:

  One of the most important requirements of the full-electronic tap-changer is its capability to stand against over-current and over-voltage during the network faults. So it is essential to design an appropriate protection system [1].

- Evaluation of the reliability of the full-electronic tap-changer system and comparison with the reliability of the mechanical tap-changers and optimization of power part based on the reliability criterion.
- Design of the tap-changer for high power transformer (distribution transformers).
- Optimization of the switches configuration in the full-electronic tap-changer assuming pre-defined structure for taps winding.
- Full analysis of the arc in the mechanical tap-changers, design and optimization of the hybrid tap-changer for reduction of the arc.
- Extending the designs to three-phase case.
- Designing a commutation method for bi-directional switches without current and voltage sensors.
- Designing a digital controller for the full-electronic tap-changer.
- Designing the full electronic tap-changer with soft switching for application of pulsed-width modulation.
- Comparison of the full-electronic tap-changer with PWM and AC/AC converters [2–10].
- Designing a fast detection method for the sag in order to use it in the control system of the full-electronic tap-changer.

References

1. Bauer P, de Haan SWH (1997) Protective device for electronic tap-changer for distribution transformers, EPE 97, Sept 1997, pp 4/282–4/287
2. Hietpas SM, Naden M (2000) Automatic voltage regulator using an AC voltage-voltage converter. IEEE Trans Ind Appl 36(1):33–38
3. Venkataramanan G, Johnson BK, Sundaram A (1996) An AC-AC power converter for custom power applications. IEEE Trans Power Deliv 11(3):1666–1671
4. Ronan ER, Sudhoff SD, Glover F, Galloway DL (2002) A power electronic-based distribution transformer. IEEE Trans Power Deliv 17(2):537–543
5. Johnason BK, Venkataramanan G (1998) A hybrid solid state phase shifter using PWM AC converters. IEEE Trans Power Deliv 13(4):1316–1321
6. Venkataramanan G, Johnson B (1997) A pulse width modulation power line conditioner for sensitive load centers. IEEE Trans Power Deliv 12(2):844–849
7. Hietpas SM, Pecen R (1998) Simulation of a three-phase AC-AC boost converter to compensate for voltage sags, Rural Electric Power Conference Apr 1998, pp B4/1–B4/7
8. Kwon BH, Youm JH, Choi JH (1999) Automatic voltage regulator with fast dynamic speed. IEE Proc Electr Power Appl 146(2):201–207
9. Venkataramanan G (1997) A family of PWM converters for three phase AC power conditioning, PESC 1997, pp 572–577
10. Lopes LAC, Joos G, Ooi BT (1997) A PWM quadrature booster phase shifter for AC power transmission. IEEE Trans Power Electron 12(1):138–144

MIX
Papier aus verantwortungsvollen Quellen
Paper from responsible sources
FSC® C105338

If you have any concerns about our products,
you can contact us on
ProductSafety@springernature.com

In case Publisher is established outside the EU,
the EU authorized representative is:
Springer Nature Customer Service Center GmbH
Europaplatz 3, 69115 Heidelberg, Germany

Printed by Libri Plureos GmbH
in Hamburg, Germany